Roman Petres

Crystalline Silicon Surface Passivation by Amorphous Silicon Compounds

Roman Petres

Crystalline Silicon Surface Passivation by Amorphous Silicon Compounds

Modeling, experiments, solar cells and modules

Südwestdeutscher Verlag für Hochschulschriften

Impressum/Imprint (nur für Deutschland/only for Germany)
Bibliografische Information der Deutschen Nationalbibliothek: Die Deutsche Nationalbibliothek verzeichnet diese Publikation in der Deutschen Nationalbibliografie; detaillierte bibliografische Daten sind im Internet über http://dnb.d-nb.de abrufbar.

Alle in diesem Buch genannten Marken und Produktnamen unterliegen warenzeichen-, marken- oder patentrechtlichem Schutz bzw. sind Warenzeichen oder eingetragene Warenzeichen der jeweiligen Inhaber. Die Wiedergabe von Marken, Produktnamen, Gebrauchsnamen, Handelsnamen, Warenbezeichnungen u.s.w. in diesem Werk berechtigt auch ohne besondere Kennzeichnung nicht zu der Annahme, dass solche Namen im Sinne der Warenzeichen- und Markenschutzgesetzgebung als frei zu betrachten wären und daher von jedermann benutzt werden dürften.

Coverbild: www.ingimage.com

Verlag: Südwestdeutscher Verlag für Hochschulschriften GmbH & Co. KG
Dudweiler Landstr. 99, 66123 Saarbrücken, Deutschland
Telefon +49 681 37 20 271-1, Telefax +49 681 37 20 271-0
Email: info@svh-verlag.de

Approved by: Konstanz, Universität, Diss., 2010

Herstellung in Deutschland:
Schaltungsdienst Lange o.H.G., Berlin
Books on Demand GmbH, Norderstedt
Reha GmbH, Saarbrücken
Amazon Distribution GmbH, Leipzig
ISBN: 978-3-8381-2879-5

Imprint (only for USA, GB)
Bibliographic information published by the Deutsche Nationalbibliothek: The Deutsche Nationalbibliothek lists this publication in the Deutsche Nationalbibliografie; detailed bibliographic data are available in the Internet at http://dnb.d-nb.de.

Any brand names and product names mentioned in this book are subject to trademark, brand or patent protection and are trademarks or registered trademarks of their respective holders. The use of brand names, product names, common names, trade names, product descriptions etc. even without a particular marking in this works is in no way to be construed to mean that such names may be regarded as unrestricted in respect of trademark and brand protection legislation and could thus be used by anyone.

Cover image: www.ingimage.com

Publisher: Südwestdeutscher Verlag für Hochschulschriften GmbH & Co. KG
Dudweiler Landstr. 99, 66123 Saarbrücken, Germany
Phone +49 681 37 20 271-1, Fax +49 681 37 20 271-0
Email: info@svh-verlag.de

Printed in the U.S.A.
Printed in the U.K. by (see last page)
ISBN: 978-3-8381-2879-5

Copyright © 2011 by the author and Südwestdeutscher Verlag für Hochschulschriften GmbH & Co. KG and licensors
All rights reserved. Saarbrücken 2011

"There is no energy crisis, only a crisis of ignorance."

Richard Buckminster Fuller (1895-1983), American inventor and architect, one of the first strong supporters of renewable energy

"I'd put my money on the sun and solar energy. What a source of power! I hope we don't have to wait until oil and coal run out before we tackle that."

Thomas Alva Edison (1847-1931) in 1931

Contents

1 Introduction   **1**
   1.1 Photovoltaics: Current state and potential 1
   1.2 Motivation for this work 3
   1.3 Contribution of this work to the research field 6
   1.4 Structure of the document 8

2 Surface Passivation and Antireflection Coating   **9**
   2.1 Theory 10
      2.1.1 Surface recombination 10
      2.1.2 Surface passivation 11
      2.1.3 Antireflection coating (ARC) 12
   2.2 Characterisation of surface passivation layers 14
      2.2.1 Undiffused surfaces 14
      2.2.2 Emitter-diffused surfaces 18
      2.2.3 Photoconductance measurements 20
      2.2.4 Spectroscopic Ellipsometry 23
      2.2.5 Fourier Transform Infrared Spectroscopy (FTIR) 24

3 Fabrication of surface passivation layers   **25**
   3.1 Fabrication of surface coatings-deposition versus growth 26
   3.2 Grown films 26
      3.2.1 Thermal oxidation 27
      3.2.2 Plasma-activated oxidation 28
   3.3 Deposited films 28
      3.3.1 CVD-deposition 28
      3.3.2 PECVD-deposition 29
      3.3.3 Equipment used for this work 30

4 PECVD-Silicon Nitride   **35**
   4.1 Gas flow ratio and substrate quality dependence 36
   4.2 Boat position dependence 38
   4.3 Emitter passivation 40

4.4	Ammonia quality dependence	42
	4.4.1 Comparison of different bottle fill levels for ammonia grade N36	44
	4.4.2 Comparison of ammonia grades N50, N36 and N20	49
	4.4.3 Module level testing	52
	4.4.4 Conclusions	57

5 PECVD-Silicon Carbide and Silicon Carbonitride   61

5.1	PECVD-SiC$_x$ for surface passivation	62
5.2	High Frequency Direct Plasma	62
	5.2.1 p$^+$-Si Passivation	62
	5.2.2 p- and n-type Si passivation	62
	5.2.3 Etching behavior	69
5.3	Low Frequency Direct Plasma	69
	5.3.1 First experiments	69
	5.3.2 SiC$_x$: surface passivation dependence on deposition parameters	71
	5.3.3 SiC$_x$N$_y$: surface passivation dependence on deposition parameters	72
	5.3.4 p$^+$-passivation	75
5.4	DoE on gas flow ratio, power, chamber pressure and temperature	76
	5.4.1 General results	78
	5.4.2 Gas flow ratio and plasma power dependence	78
	5.4.3 p$^+$-passivation	81
	5.4.4 Preplasma dependence	82
5.5	FTIR-study of SiC$_x$ layers from low-frequency PECVD	84
5.6	Comparison of a-SiC$_x$:H to a-SiN$_x$:H	88

Bibliography   91

Summary   95

Zusammenfassung   97

Publications   101

Acknowledgements   103

1

Introduction

1.1 Photovoltaics: Current state and potential

Among the known energy sources that can be employed today, there is no other one being as abundant as our sun. The amount of solar energy reaching our earth within one hour equals the total annual energy need of all of mankind, taking into account both heat and electricity.

Technologies to use the sun's energy directly for satisfying those energy needs are readily available. The only thing left is thus their large-scale implementation, which can only result from a large-scale desire for them by the people. Besides the big ecological advantages of solar energy, the price per kWh to date still seems to be the key factor to create this desire. The dwindling supplies of coal, oil, natural gas and uranium are leading to steadily and in recent years also rapidly increasing costs of using them. Solar energy, on the other hand, is becoming ever cheaper. Solar heat is already cost-competitive or close to becoming it, depending on the individual countries' supply and price situation of other, usually non-renewable sources.

In most areas of the world that are not connected to a grid and in sunny regions like Spain, southern Italy or Hawaii, even solar electricity generation by photovoltaics (PV), the direct conversion of sunlight into electricity, has already approached or surpassed the threshold into the so-called grid-parity price region. This means that the price per kWh, assuming a moderate lifespan of 20 years for the PV system, is in the same region as the one a consumer has to pay for electricity generated from conventional, non-renewable sources. The grid-parity is an important benchmark on the way of solar energy to become one of the dominant energy sources on earth.

In less sunny regions like central and northern Europe or countries with large subsidies for non-renewable sources like Australia, further efforts are necessary to reduce the price per kWh by at least a factor of two, or a module price below €1 per watt-peak (W_p). On-going up-scaling of factories and the current global economy crisis have recently brought down module prices close to grid-parity in Germany already with €1.8 per W_p for European and €1.4 per W_p for Chinese producers as of May 2010, but there is still a lot of potential for further improvements*.

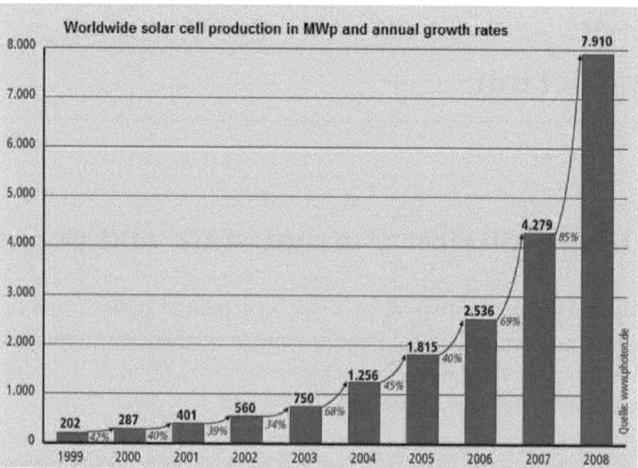

Figure 1.1: *Growth of the PV industry from 2000-2008. In 2009, the total production reached 12.4 GWp [Photon10], corresponding to 56% growth as compared to 2008.*

What are the expectations for the near future regarding PV cost? Not taking into account further improvements in cell efficiency while decreasing Si consumption and using lower purity grade and thus cheaper Si feedstock, the learning curve of the last decades, see figure 1.2, already suggests another 50% of price reduction, among with a 10-fold increase in production capacity. This is likely to happen within the next 5 years, assuming an average annual production growth rate of 35%, which was in fact consistently exceeded within the last ten years (2000-2009) with a compound annual growth rate (CAGR) of 49,7% [Solbuz10,Solsrv09]. The financial crisis has little effect on the growth, as the annual production has exceeded 10 GW_p in 2009.

---

*Not taking into account interest rates and assuming currently available total system costs of below €3.0 per W_p, this translates into generation costs of as low as €0.15 per kWh in southern Germany when assuming a conservative plant lifetime of 20 years. This is already below the current household consumer price for conventionally generated electricity in Germany of about €0.20 per kWh that does not include monthly grid connection fees, which can result in an effective price of above €0.30 for a household with below-average electricity comsumption (600 kWh/year).

1.2. Motivation for this work 3

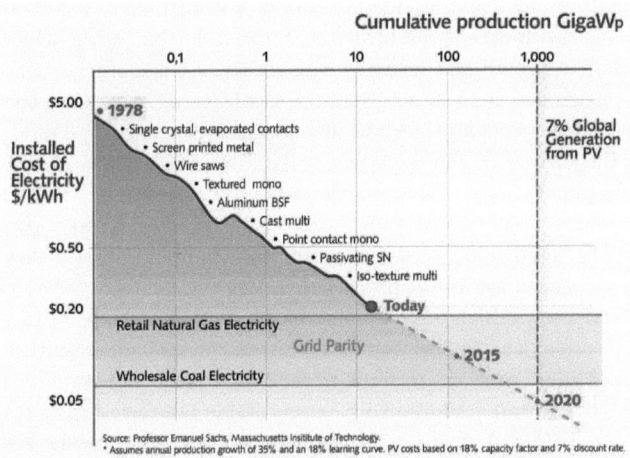

Figure 1.2: *Learning curve for the price per kWh of electricity from Photovoltaics. Over the last 30 years, PV cost have been decreasing by an average 10% per year, while production has gone up by an average 25% per year. In the last 10 years, the compound annual growth rate was twice as high. Projecting an average annual growth of 35% into the next 12 years and assuming the same 10% annual cost decrease, PV will be cheaper than coal generated electricity already by 2020 [Sachs08].*

1.2 Motivation for this work

Considering crystalline silicon PV which still dominates the market by over 80% [Solbuz10], the major path to further decrease PV cost at the cell level is to produce solar cells with higher efficiencies while at the same time lowering the silicon cost by using cheaper material and/or consuming less silicon by using thinner wafers and/or developing wafering technologies that are less wasteful than wire-sawing.

While alternative technologies that produce wafers directly from molten silicon such as edge-defined film-fed growth (EFG) and ribbon growth on substrate (RGS) in laboratories have achieved efficiencies similar to current industrial values for solar cells made with wire-cut wafers, the lower costs per wafer are not yet sufficient to compensate for the efficiency drop observed in industrial production. Other issues are the rough surfaces (RGS) or the physical limitation to special or smaller wafer formats (EFG).

Current industrial wafer thicknesses are in the range of 150-200 μm, compared to 300 μm by and still after the year 2000, and 400 μm in the 1980's. This progress towards thinner wafers was made possible by improvements in the wire-sawing technique and production machines being able to handle the fragile thinner wafers with acceptable breakage. Simulations balancing achievable efficiency (assuming good light trapping and surface passivation)show an optimum for wafer thicknesses of 40-90 μm [Kerr02,Kerr03,Geer04] depending on substrate doping, with the optimum thickness increasing with resistivity. For 1 Ωcm, the optimum thickness is 55 μm according to [Kerr03]. Such thin wafers are flexible and not fragile anymore. However, the necessary adaptations in the solar cell manufacturing process to such thin wafers will be challenging to implement.

So far, the trend towards even thinner wafers is limited by the fact that the percentage of silicon lost in the process of cutting the wafers by wire-sawing (the so-called kerf-loss) is higher for thinner wafers as the thickness of the wire can hardly be reduced any further than the current 150-200 μm (resulting in up to 50% of the initial Si lost as kerf-loss already for current industrial wafer thicknesses of 150-200 μm). Additionally, unacceptable breakage rates of below 180 μm thick wafers (mainly multicrystalline ones) occur in current solar cell/module manufacturing machines.

Recently, an alternative wafering method to wire-sawing has been demonstrated [Henley09] that can produce mono-crystalline wafers as thin as 20 μm almost without the kerf loss associated with wire-sawing. This technology is based on implanting hydrogen ions with a well defined energy and thus penetration depth (e.g. 20 μm) into a brick of silicon. Subsequently, the wafer is separated from the brick by applying mechanical tension at one of the brick's sides, creating a well-defined crack along the plane of implanted hydrogen. According to the producer of the system, it should be able to well compete with wire-sawing in terms of costs per wafer. However, even when avoiding the handling issues to be solved with 20-50 μm thin wafers by cleaving 150 μm wafers in this way, the fact that the method results preferably in wafers with (111)-surfaces is a potential drawback for the integration into existing production facilities so far, as it renders the currently most common texturization method for mono-crystalline wafers (wet-chemical random pyramid texturization by preferential etching of the (100) oriented surface) impossible. However, apart from plasma-etching, a recently developed relatively simple wet-chemical etching approach [Gabor10] that is isotropical (i.e. independent of crystal orientation) yields similar light capturing quality and might be a viable alternative.

To date, the ion implantation method has been demonstrated only on mono-crystalline bricks produced with the Czochralski (Cz) or Float Zone (FZ) method. This does not have to be a disadvantage, as for multi-crystalline bricks, the mechanical stability of wafers below 100 μm thickness would be a serious issue, thus possibly favoring mono-crystalline wafer-based cell concepts in the future in case the wafer thickness trend will continue towards below 100 μm. This will depend on the market price of Si, the development of technologies to safely handle wafers thinner than 100 μm in an automated cell line and the availability of cost-effective processes delivering the required qualities of light-trapping and

1.2. Motivation for this work

especially electronic surface passivation.

The influence of surface passivation quality on solar cell efficiency is increasing with decreasing wafer thickness. Figure 1.3 shows the PC1D-simulated efficiency gain for thinner cells, comparing 1 Ωcm high-quality material (effective diffusion length L_{eff}=1000 µm, i.e. bulk carrier lifetime τ_{bulk}=400 µs) to 0.3 Ωcm solar grade material (L_{eff}=200 µm, i.e. τ_{bulk}=20 µs).

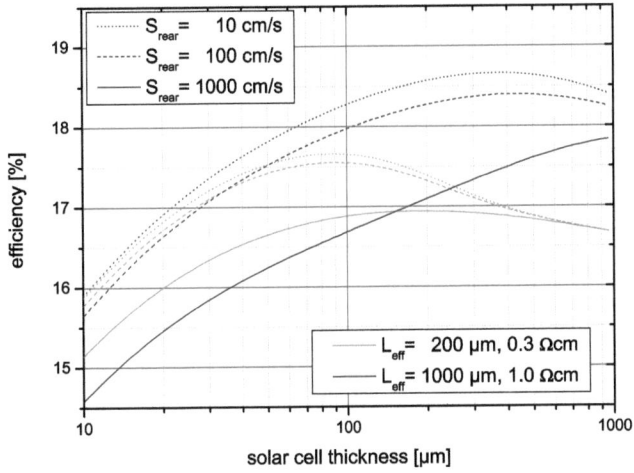

Figure 1.3: *Increasing importance of surface passivation with decreasing cell thickness and higher efficiencies with decreasing thickness for lower quality and thus cheaper Si material. With the already realized value of $S_{rear} = 100 cm/s$, a 30 µm solar cell would show the same performance with either material.*

The simulated cell design features a textured front side and single-layer antireflection coating, resulting in 5 % overall reflectance, screen-printed contacts with a passivated open rear contact and a passivated homogeneous 80 $\Omega/\square$ front side emitter with a peak doping concentration of $1 \cdot 10^{20}$ cm^{-3} and a front surface recombination velocity of 10^4 cm/s. Solar cell efficiencies are calculated for three different rear surface recombination velocities of 10, 100 and 1000 cm/s. While 1000 cm/s can be easily reached in reality by a mediocre Aluminium back surface field interrupted by Ag-pads for stringing, and 100 cm/s by e.g. laser-fired contacts on 0.5 Ωcm material [Grohe03], 10 cm/s including the metallized areas have not yet been reported (for the dielectrically passivated areas only, 10 cm/s and below are possible). If this target can be reached at all with solar cells featuring direct Si-metal contacts, it might only follow from a deeper understanding of the physics at the Si-metal including Si-dielectric interfaces. However, the largest effiency leap clearly comes already with the transition from a full area Al-metallization and Al-BSF to dielectric rear-side passivation and local contacts and BSF.

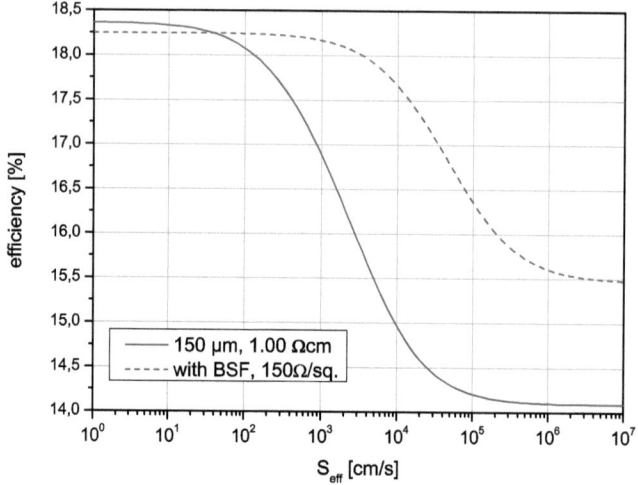

Figure 1.4: *A diffused back surface field (BSF) decreases the dependence of cell efficiency on rear surface passivation quality, but also limits the maximum efficieny for a given structure.*

As shown in figure 1.4, highest efficiencies can be achieved with excellent dielectric rear side passivation without an underlying diffused BSF, but a diffused BSF decreases the dependence of cell efficiency on rear surface passivation quality. Thus, it seems reasonable to always apply at least a lightly diffused BSF and thus greatly broaden the processing window, as S_{rear} can be up to two orders of magnitude higher with a BSF, without considerable losses in cell efficiency.

1.3 Contribution of this work to the research field

This PhD thesis contributes to the research field of dielectric surface passivation layers with the following elements:

- The observed surface passivation quality by SiN_x (chapter 4) is, to the knowledge of the author, be the highest published for a low-frequency

1.3. Contribution of this work to the research field 7

PECVD system so far and is only slightly below the best published values for high-frequency PECVD [Kerr03]. This is in contradiction to previous publications reporting inferior surface passivation quality of low-frequency PECVD systems due to ion-bombardment induced surface damage. A likely explanation is offered for the similarly high passivation quality of our low-frequency PECVD layers as compared to high-frequency PECVD layers.

- The influence of the gas purity of the ammonia used for depositions of silicon nitride was investigated for the first time.

- For the first time, studies are presented on the influence of several deposition parameters for plasma enhanced chemical vapor deposition (PECVD) on the electronic surface passivation and optical properties of silicon carbide (a-SiC$_x$:H) and silicon carbonitride(a-SiC$_x$N$_y$:H) layers in Chapter 5 using a low-frequency PECVD system for depositions. The DoE study presented in chapter 5.4 appears to cover a larger parameter space than previously published studies on SiC$_x$.

- Two different types of industrial PECVD reactors were used: a high-frequency (13.56 MHz) and a low-frequency (40 kHz) direct plasma PECVD reactor from different manufacturers of production equipment. Thus, the results can be directly implemented in solar cell production lines using the same equipment. The high-frequency reactor was only used for a-SiC$_x$:H, while a-SiN$_x$:H, a-SiC$_x$:H, a-SiC$_x$N$_y$:H and a-SiO$_x$N$_y$:H were deposited with the low-frequency system. To the knowledge of the author, this is the first time that results of low- and high-frequency direct plasma PECVD were compared for a-SiC$_x$:H.

Additionally, in the frame of this work, a dielectric layer stack of PECVD-SiO$_x$N$_y$/SiN$_x$ was developed that fulfills all necessary requirements to allow for a cell efficiency improvement of 0.5% compared to a full area Al-BSF cell (calculated from measured V_{oc} and I_{sc} improvements, as the fill factor was limited by the material), while using a processing sequence of comparable simplicity and without the need for any additional process equipment other than an additional process gas line for the PECVD system.

While the actual solar cell characteristics as well as process sequence and dielectric film deposition parameters for this stack may currently unfortunately not be published due to intellectual property protection of the company it was developed for, the surface passivation performance which is the highest presented at standard solar cell working conditions in this work (1 sun illumination) is shown in chapter 5.4.

1.4 Structure of the document

Chapter 2 gives an introduction to the theoretical background of surface recombination, surface passivation and antireflection coating, and briefly explains the working principles and methodology of the characterization instruments used in this work.

Chapter 3 gives an overview of different fabrication methods for dielectric surface coatings (deposition and growth) with a focus on PECVD and here mainly the low-frequency system used for most experiments in this work.

Chapter 4 describes the results of experiments carried out with silicon nitride layers from low-frequency PECVD, comparing them to previously published studies. Besides the influence of ammonia to silane gas flow ratio, the wafer position in the boat during deposition and etched-back emitters of various etching depth and thus sheet resistivity, the influence of gas purity of the ammonia used for the depostions is investigated.

Chapter 5 describes the results of the experimental investigations on silicon carbide and carbonitride, investigating the influence of precursor gas flow ratios, deposition temperature, chamber pressure and plasma power. Comparison to other low-frequency silicon carbide or carbonitride studies in literature was not possible due to the lack of previous experiments with such equipment.

2

Surface Passivation and Antireflection Coating

Abstract

This chapter gives an overview of the theoretical bases of surface passivation and antireflection coating and describes the methods and equipment used to characterize the layers created in this work. While surface passivation is quantified by the effective surface recombination velocity S_{eff}, this parameter cannot be measured directly. Instead, the lifetime measurements by QSSPC and µPCD carried out for this work give the effective minority carrier lifetime τ_{eff}. With certain simplifying assumptions, an upper limit for S_{eff} can be calculated solely from τ_{eff} and the sample thickness. As shown in chapter 2.2.1, the error resulting from this simplified approach often found in literature is not negligible for good surface passivation layers, but acceptable in practice as the focus is on comparing different passivation layers.

While the µPCD was applied to obtain spatially resolved lifetime maps of the entire sample, the QSSPC was subsequently used to determine absolute values of the best areas that can be compared with the literature, as QSSPC is the established standard in c-Si photovoltaics.

The refractive index and thickness of the investigated dielectric films were measured by spectroscopic ellipsometry, and the chemical composition was analyzed by Fourier-Transformed Infrared Spectroscopy (FTIR) to investigate relations with the surface passivation and optical properties.

2.1 Theory

2.1.1 Surface recombination

At a crystalline semiconductor surface, the crystal lattice is completely lost. That means the atoms at the surface will usually have non-saturated (also called dangling) bonds which cause a high density of interface states D_{it} per unit of area (usually given in cm^2) within the forbidden bandgap of the semiconductor which act as potential recombination sites. Further possible sources of interface states at the surface are impurities like organic residues and metals or process-induced additional crystal defects, e.g. from chemical or mechanical etching. In analogy to defect-induced bulk recombination, surface recombination via a defect at an energy level E_t can be described by the Shockley-Read-Hall theory, placing the electron and hole surface carrier densities n_s and p_s [cm^{-3}] instead of the bulk densities n and p and the density of this defect per unit area N_{ts} instead of the defect density N_t per unit volume, thus obtaining a surface recombination rate R_s as follows:

$$R_s = \frac{v_{th} N_{ts}(n_s p_s - n_i^2)}{\frac{n_s+n_1}{\sigma_p} + \frac{p_s+p_1}{\sigma_n}} \qquad (2.1)$$

where n_i is the intrinsic excess carrier concentration [cm^{-3}], v_{th} is the thermal velocity of charge carriers (10^7 cm·s^{-1} in c-Si at 300K [Martin03]), $n_1 = n_i \cdot e^{\frac{E_t-E_i}{kT}}$, $p_1 = n_i \cdot e^{\frac{E_i-E_t}{kT}}$ and $\sigma_n(\sigma_n)$ are the capture cross section [cm^2] of the defect for electrons(holes).

This equation is often presented as:

$$R_s = \frac{n_s p_s - n_i^2}{\frac{n_s+n_1}{S_{p0}} + \frac{p_s+p_1}{S_{n0}}} \qquad (2.2)$$

$$S_{n0} \equiv v_{th} D_{it} \sigma_n \qquad (2.3)$$

$$S_{p0} \equiv v_{th} D_{it} \sigma_p \qquad (2.4)$$

where S_{n0} and S_{p0} are the so-called fundamental recombination velocities of electrons and holes, respectively. In analogy to bulk recombination, where the recombination rate R is of the dimension [$cm^{-3}s^{-1}$] and a recombination lifetime τ of excess carriers $\Delta n = \Delta p$ is defined via

$$R = \frac{\Delta n}{\tau} \qquad (2.5)$$

the dimension [$cm^{-2}s^{-1}$] of R_s suggests the definition of a surface recombination velocity S [cm/s] via

$$R_s = S \cdot \Delta n_s \qquad (2.6)$$

with Δn_s [cm^{-3}] being the excess minority carrier density at the surface (in case of p-type material) that would equal Δp_s in absence of an electric field.

2.1. Theory

An important difference between bulk recombination via defects and surface recombination is now the point that an electric field is usually found at the semiconductor surface. In this case, Δn_s is far away from Δp_s since the electric field creates large differences between n_s and p_s. It thus makes sense to define an effective surface recombination velocity S_{eff} as follows:

$$S_{eff} = \frac{R_s}{\Delta n} \qquad (2.7)$$

where $\Delta n = \Delta p$ is the excess minority carrier density at the limit of the space charge region which is created at the surface and which equals the bulk excess carrier density. In contrast to Δn_s, Δn can easily be measured and controlled by changing the illumination level.

2.1.2 Surface passivation

Surface passivation in the electronic sense means avoiding the recombination of minority carriers at the semiconductor surface. While the term "surface passivation" is also used in chemistry to describe the act of rendering the surface of a certain substance chemically inert, it is solely used to describe electronic surface passivation in this work.

As surface recombination of electron-hole pairs is taking place at surface defects, it can consequently be reduced by either rendering surface states inactive or keeping one kind of carriers from reaching the surface, as both species are needed for recombination to occur.

Mathematically, these two ways can be deducted from equation 2.2:

- *Reduction of the fundamental recombination velocities of electrons and holes, S_{n0} and S_{p0}:* This can be achieved by lowering the interface state density D_{it}. As interface states result from dangling bonds, these bonds thus have to be saturated. This can be achieved i) either by deposition/growth conditions of a surface layer that allow sufficient time (defining a maximum deposition rate) and energy (defining a minimum temperature or plasma energy density) for atoms to reach an energetically optimal location which a dangling bond constitutes, or ii) by a post-deposition treatment like the common "firing" step in Si solar cell production that enables hydrogen atoms from a H-containing surface dielectric like PECVD-deposited SiN_x to diffuse and stick to these dangling bonds.

- *Reduction of the surface concentration of electrons (n_s) or holes (p_s):* This can be achieved by means of an electric field close to the surface that repels either minority or majority carriers. The former case is called accumulation, the latter inversion. Because of its similarity to an uncontacted emitter, the latter is also named "floating junction".

The electric field needed to achieve surface depletion of one kind of carriers can be provided in two ways:

- "Integrated" into the emitter or into the rear side (back-surface field, BSF) of the solar cell by means of a gradient in dopant concentration, with increasing dopant density towards the surface. Conveniently, such a gradient occurs intrinsically when creating the emitter or BSF via phosphorus or boron diffusion or aluminium-alloying.

- Via fixed charges within a dielectric layer grown or deposited on the Si surface which create a band bending near the interface inside the Si and thus induce a near-surface charge inside the silicon, thus also creating an electric field gradient near the surface.

Such a dielectric layer should further have the following beneficial properties:

- It can saturate a large majority of the Si surface states, either directly during deposition or after activation by a high-temperature step like the metal contact co-firing.

- It is transparent for the part of the EM radiation spectrum that the solar cell is sensitive to in a module (in the case of Si encapsulated with low-iron glass, this corresponds to an energy gap of at least 4 eV)

- It has a homogeneous thickness

- It has a refractive index that allows for good antireflection coating.

2.1.3 Antireflection coating (ARC)

Reduction of reflection at a silicon surface can be achieved by two basic effects:

- Reduction of the difference in refractive indices at the air/silicon interface for non-encapsulated solar cells and the encapsulant/silicon interface for encapsulated cells, respectively.

- Destructive interference in between incident and reflected light waves.

To describe these two mechanisms in detail, a planar, polarized electromagnetic wave can be defined. A phase change δ of this EM wave propagated in a medium with refractive index n > 1

$$\delta = \frac{2\pi}{\lambda_0} nd \qquad (2.8)$$

is directly proportional to the optical path of the light wave, i.e. the product of the layer thickness d and the refractive index n, and is inversely proportional to the wavelength in vacuum λ_0. Assuming normally incident light and a coating on Si with an optical thickness of $n_1 d = \lambda_0/4$, Fresnel's equations yield the reflection

2.1. Theory

$$R = \left(\frac{n_0 n_s - n_1^2}{n_0 n_s + n_1^2}\right)^2 \qquad (2.9)$$

which depends on the refractive index of Si, n_s, refractive index of the coating film n_1, and the refractive index of the ambient n_0.

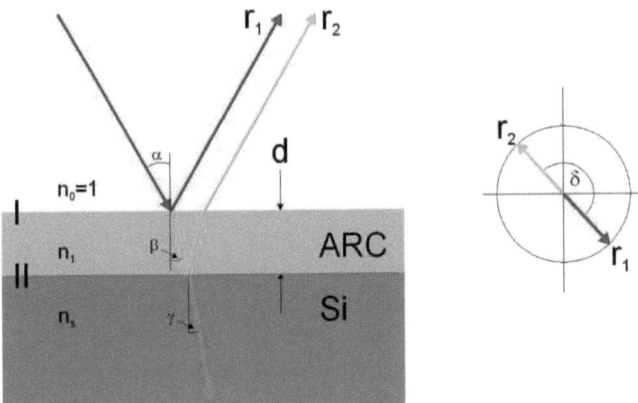

Figure 2.1: *Schematic of single layer antireflection coating on a substrate with refractive index n_s and in an ambient with refractive index n_0. Reduction of reflectance occurs only for a narrow wavelenght range, in which a phase offset of $\delta = 180°$ in between the first (red) and second (yellow) reflected wave of equal amplitude cause maximum destructive interference [Per03].*

Within a single layer ARC (see fig. 2.1), a phase change of $\delta = 180°$ and equal amplitudes of two waves, one reflected at the upper, and the other at the lower interface of the ARC, cause a maximum destructive interference

$$\frac{n_0 - n_1}{n_0 + n_1} \approx \frac{n_1 - n_{Si}}{n_1 + n_{Si}} \rightarrow \frac{n_0}{n_1} = \frac{n_1}{n_{Si}} \qquad (2.10)$$

This yields the optimum refractive index n_1 of a single layer ARC and its optimum thickness d from the following equations:

$$n_1 = \sqrt{n_0 n_{Si}}; \qquad d = \frac{\lambda_0}{4n_1} \qquad (2.11)$$

where n_{Si} is the refractive index of silicon ($n_{Si} = 3.87$ at $\lambda = 632.8$ nm) and n_0 is the refractive index of the ambient. In the case of non-encapsulated cells, $n_0 = n_{air} = 1$, and for cells encapsulated under EVA or silicone, $n_0 \approx 1.5$.

A single layer ARC provides a large reduction of reflection losses, but localized around the specific wavelength for which the film was designed. To achieve a further decrease in reflection, a multi-layer ARC must be configured. The

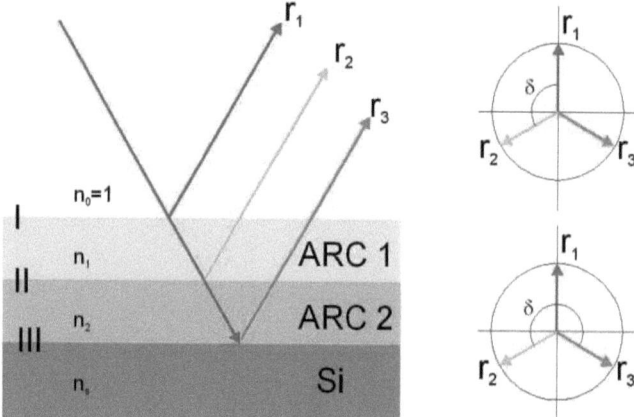

Figure 2.2: *Schematic of a double layer antireflection coating on a substrate with refractive index n_s and in an ambient with refractive index n_0. As destructive interference can occur in between the first (red), second (yellow) and third (green) reflected ray, reflection minima can be found not only for a phase offset of $\delta = 180°$, but also at $\delta = 120°$ and $\delta = 240°$, corresponding to wavelengths λ_0, $3/4\lambda_0$ and $3/2\lambda_0$ [Per03].*

refractive indices of a double layer ARC can be determined in analogy to that of a single layer ARC:

$$\frac{n_0}{n_1} = \frac{n_1}{n_2} = \frac{n_2}{n_{Si}} \rightarrow \left(\frac{n_2}{n_1}\right)^2 = \frac{n_{Si}}{n_0} \qquad (2.12)$$

where n_1 and n_2 are now the refractive indices of the upper and lower ARC layer, respectively. In the phase diagram of fig. 2.2, it can be observed that a double layer ARC provides a reflection minimum not only for $\delta = 180°$, but also for $\delta = 120°$ and $\delta = 240°$, which corresponds to wavelengths of λ_0, $3/4\lambda_0$ and $3/2\lambda_0$.

2.2 Characterisation of surface passivation layers

2.2.1 Undiffused surfaces

The available methods to measure excess minority carrier lifetimes of a sample allow for measuring effective lifetimes τ_{eff} only. These contain by definition the contributions of all recombination processes possible within a sample, which is

2.2. Characterisation of surface passivation layers

mathematically expressed by the equation

$$\sum_i R_i = \frac{\Delta n_{av} \cdot W}{\tau_{eff}} \qquad (2.13)$$

wherein the average excess minority carrier density Δn_{av} itself is defined via

$$\Delta n_{av} = \frac{1}{W} \int_{-W/2}^{W/2} \Delta n(x) dx \qquad (2.14)$$

where W denotes the thickness of the c-Si sample. Depending on the wavelenght of the incident light generating the excess carriers, different profiles of $\Delta n(x)$ result from the absorption coefficient $\alpha(\lambda)$.

Now, the wish is to separate the contributions of the bulk and the surface in order to determine the effective surface recombination velocity S_{eff}.

This is simplified by the following conditions:

1. Length y and width z of the sample greatly exceed its thickness x, and the diffusion lenght of the excess minority carriers is many times smaller than the irradiated area, in which the irradiation intensity is independent of y and z. Thus, a one-dimensional approach is sufficient.

2. The bulk lifetime τ_{bulk} is constant within the wafer and independent of the injection level.

3. Both surfaces have the same S_{eff}, which implies a symmetrical sample structure.

4. The Photo-Generation rate G_{ext} is the same everywhere within the wafer. This can only be realized if the mean penetration depth of the indicent light is several times larger than the wafer thickness x. For crystalline Si, that requires infrared light of wavelengths $\lambda > 1\mu m$. Under these conditions, the profile of $\Delta n(x)$ becomes symmetrical, meaning $\Delta n(-W/2) = \Delta n(W/2)$

The definition (2.14) can be expressed accordingly as

$$\sum_i R_i = \int_{-W/2}^{W/2} R_{bulk} dx + R_{s,front} + R_{s,back} \qquad (2.15)$$

Applying the recombination rate definitions from eq.(2.5) and (2.6) yields

$$\sum_i R_i = \frac{\int_{-W/2}^{W/2} \Delta n(x) dx}{\tau_{bulk}} + S_{eff,front} \cdot \Delta n(W/2) + S_{eff,back} \cdot \Delta n(-W/2) \qquad (2.16)$$

Because it is assumed that both surfaces are identical (thus $S_{eff,front} = S_{eff,back}$) and the profile of $\Delta n(x)$ is symmetrical, eq. (2.16) is simplified together with the definition of Δn_{av} to

$$\frac{\Delta n_{av} \cdot W}{\tau_{eff}} = \frac{\Delta n_{av} \cdot W}{\tau_{bulk}} + 2 S_{eff} \cdot \Delta n(W/2) \qquad (2.17)$$

τ_{eff} can be isolated:

$$\frac{1}{\tau_{eff}} = \frac{1}{\tau_{bulk}} + 2 \frac{S_{eff}}{W} \cdot \frac{\Delta n(W/2)}{\Delta n_{av}} \qquad (2.18)$$

Now we have almost a complete way of determining S_{eff} from τ_{eff}. Two variables remain: τ_{bulk} and the ratio of $\Delta n(W/2)$ and Δn_{av}.

As for the relationship of $\Delta n(W/2)$ and Δn_{av}, the general steady-state solution, given e.g. in [Brody03], is complex:

$$\tau_{eff} = \frac{A \cdot L \cdot sinh(W/2L) + \sum_\lambda \dfrac{sinh[\alpha(\lambda)W/2]/\alpha(\lambda)}{1/\tau_{bulk} - D \cdot \alpha(\lambda)^2}}{\sum_\lambda sinh[\alpha(\lambda)W/2]/\alpha(\lambda)} \qquad (2.19)$$

with the constant A itself given by

$$A = \frac{\dfrac{D \cdot \alpha(\lambda) sinh[\alpha(\lambda)W/2] + S_{eff} cosh[\alpha(\lambda)W/2]}{1/\tau_{bulk} - D \cdot \alpha(\lambda)^2}}{S_{eff} cosh(W/2L) + (D/L) sinh(W/2L)} \qquad (2.20)$$

In most practical cases, this complex equation can be replaced by the approximation that the excess minority carrier density at the surface is similar to that in the bulk, i.e. $\Delta n(W/2) = \Delta n_{av}$ and thus

$$\frac{1}{\tau_{eff}} = \frac{1}{\tau_{bulk}} + \frac{2 S_{eff}}{W} \qquad (2.21)$$

This holds true (meaning the error compared to equation 2.19 is below 10%) for values of $S_{eff} < 1000$ cm/s, corresponding to an effective lifetime of $\tau_{eff} > 10$ μs for $\tau_{bulk} > 1$ ms at a wafer thickness of 200 μm, which is the case for the Cz- and FZ-Si lifetime samples in this work. Figure 2.3 is extracted from [Martin03] and shows the dependence of the excess minority carrier density profile in between the two wafer surfaces on S_{eff}, assuming a bulk lifetime of 4 ms.

On the other hand, in the case of very high S_{eff} values, eq. (2.19) tends towards

$$\frac{1}{\tau_{eff}} = \frac{1}{\tau_{bulk}} + \frac{\pi^2}{W^2} D \qquad (2.22)$$

In this case, the term related to τ_{bulk} is usually negligible in front of the surface recombination. Then, the minimum measurable value of τ_{eff} can be defined as

2.2. Characterisation of surface passivation layers

$$\tau_{eff,min} = \frac{W^2}{\pi^2} \frac{1}{D} \quad (2.23)$$

$\tau_{eff,min}$ relates to the time needed by the photogenerated minority carriers for diffusing from the bulk to the high-recombination surfaces. Thus, the lifetime is only determined by the minority carrier diffusion constant and the thickness of the sample. For 4 Ωcm p-type wafers of 200 μm thickness which were used for many of the lifetime samples in this work, this gives a minimum effective lifetime of 1.1 μs. The corresponding maximum $S_{eff} \approx 10^7$ cm/s equals the thermal velocity of the minority charge carriers

$$v_{th} = \sqrt{\frac{2k_BT}{m}} \quad (2.24)$$

at T $\approx$ 300°K.

A lower limit for τ_{bulk} can be determined by applying a chemical surface passivation by iodine-ethanol to the uncoated and undiffused sample wafer which can yield $S_{eff} < 10$ cm/s.

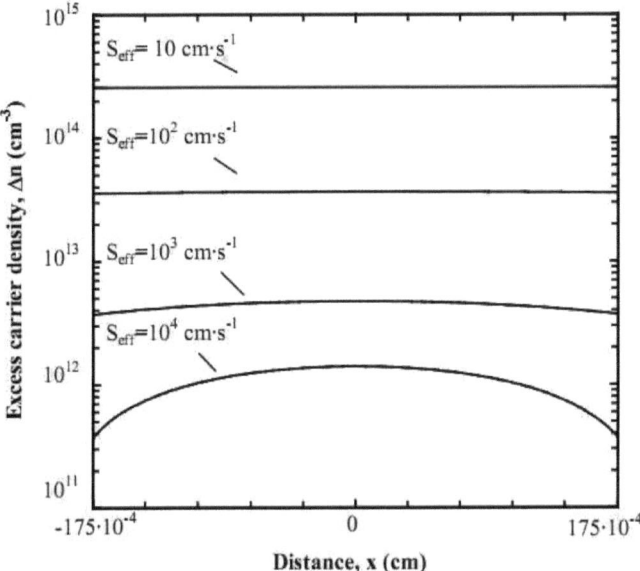

Figure 2.3: *Simulated excess minority carrier density profiles for S_{eff} ranging from 10 to 10^4 cm/s. The profiles have been simulated using PC1D (p-type Si, 3.3 $\Omega\cdot$cm, $\tau_{bulk} = 4$ ms, $\lambda_{illum} = 1140$ nm, generation rate $G_{ext} = 2.12 \cdot 10^{17} cm^{-3}$)(taken from [Martin03])*

A reasonable upper limit for τ_{bulk} depending on the dopant concentration N_{dop}, when using effective lifetimes measured at one sun illumination (in practice corresponding to injection levels of $< 10^{16}$ cm^{-3}) or at the injection level 10^{15} cm^{-3} (then, the lifetime limit is not yet dominated by Auger recombination), seems to be the Shockley-Read-Hall limit as used by [Kerr02]:

$$\tau_{SRH} = \frac{\tau_{max}}{1 + \frac{N_{dop}}{N_{ref}}} \qquad (2.25)$$

where $N_{ref} = 1 \cdot 10^{16}$cm^{-3} is an experimentally determined constant, and τ_{max}=35 ms is determined by a curve fit of the experimental results of [Kerr02b] which so far appear to be the highest reported bulk lifetimes for both p- and n-type c-Si. For ca. 1 Ωcm c-Si material, corresponding to a doping density of 10^{16}cm^{-3} (assuming non-compensated Si), this gives an upper limit for τ_{bulk} of 17.5 ms.

To simplify calculations and as this approach is used by many authors, the bulk lifetime in eq. (2.18) is assumed to tend to infinity for the conversion of effective lifetimes into effective SRVs in this work. In that way, the calculated $S_{eff,max}$ value represents an upper limit:

$$S_{eff,max} \leq \frac{W}{2\tau_{eff}} \qquad (2.26)$$

Hence, the real S_{eff} will always be lower. As can be deducted from the formula, the error caused by this simplification will increase with decreasing S_{eff} and τ_{bulk}. The following figure 2.4 shows the relation of τ_{eff} and S_{eff} for a τ_{bulk} of 4 ms (which is a realistic value for FZ-Si of about 4 Ωcm) and also the high-quality Cz-Si used for most experiments in this work), compared to the simplified case where τ_{bulk} is assumed to tend to infinity.

The relative error made by the simplification of assuming infinite minority carrier lifetime for the crystal volume is below 10 % for effective lifetimes below 330 μs and rises to 33 % for $tau_{eff} = 1$ ms. That is not negligible, but acceptable since the focus is on comparing the passivation properties, and usually high-lifetime material of similarly high bulk-lifetimes above 1 ms is used for fabricating lifetime samples in the literature as well as in this work.

2.2.2 Emitter-diffused surfaces

In the case of highly doped emitter layers, the theory for undiffused surfaces can be adapted rather easily, especially when a symmetrically diffused structure is present as is typically the case for lifetime samples. According to the standard diode equation, the electronic quality of an emitter is characterized by the emitter saturation current density J_{0e}. As the minority carrier lifetime in highly doped regions of a symmetrically diffused sample is significantly lower than in the base

2.2. Characterisation of surface passivation layers

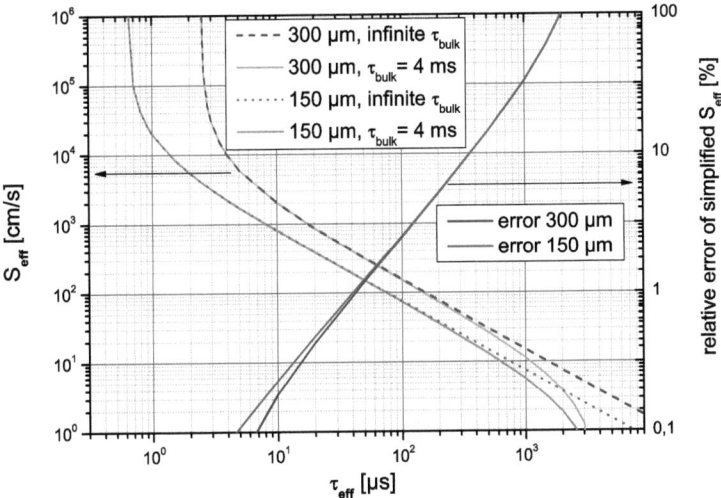

Figure 2.4: *Relation of τ_{eff} and S_{eff} for wafer thicknesses of 150 μm and 300 μm, compared for a τ_{bulk} of 4 ms and an infinite τ_{bulk}. The relative overestimation of S_{eff} made by the latter simplifying assumption is below 10% for $\tau_{eff} < 330 \mu s$ which corresponds to S_{eff} values above 20 cm/s (for 150 μm wafers) and 40 cm/s (for 300 μm wafers), respectively*

doped bulk, the recombination current into the emitters during excess minority carrier decay can be expressed quite accurately by the "quasi-static emitter approximation" [Jain81]:

$$J_{em}(V) = J_{0e}(e^{qV/kT} - 1) \qquad (2.27)$$

This equation requires low-injection conditions, which is in practice always the case for a solar cell in non-concentration operation and leads to an ideality factor of 1. Under static conditions, the voltage dependence of the minority carrier density at the edges of the depletion regions of the p/n-junctions (at $x = \pm W'/2$) is

$$\Delta n(\pm W'/2) = n_{p0}(\pm W'/2)(e^{qV/kT} - 1) \qquad (2.28)$$

Combining the two previous equations yields the following expression for the recombination current into the (in our example n$^+$) emitters:

$$J_{em}(\Delta n)|_{x=\pm W'/2} = \frac{J_{0e} N_A}{n_i^2} \Delta n \qquad (2.29)$$

The comparison with the boundary condition for an undiffused surface from eq. (2.7):

$$J_s = q \cdot R_s = q \cdot S_{eff}(\Delta n) \qquad (2.30)$$

then gives a simple relation between J_{0e} and the effective surface recombination velocity (SRV) of a n$^+$ diffused emitter on a p-type wafer under low-injection:

$$S_{eff} = \frac{J_{0e} N_A}{q n_i^2} \qquad (2.31)$$

2.2.3 Photoconductance measurements

A common way to measure effective minority carrier lifetimes is by monitoring the photoconductance (PC) of a sample. In the two techniques that were available for this work, the photoconductance decay (PCD) after pulse-like excitation by an external light source is measured with microwave reflection or inductive coupling, respectively.

As the recombination properties in the volume and at the surfaces, quantized by τ_{bulk} and S, may be strong functions of the injection level and vary over the sample, particularly in case of multicrystalline silicon, important requirements for a lifetime measurement technique are therefore to provide 1) spatial resolution and 2) absolute effective lifetimes at known injection levels.

The first requirement is met by the microwave detected photoconductance decay (μPCD), the second by the quasi-steady-state photoconductance (QSSPC) technique from which a lifetime spectrum over a large injection range is obtained. The information content of both techniques is different so that they complement each other. For this reason, lifetime samples in this work were usually characterized by both methods. Firstly, the spatial lifetime distribution was mapped with the μPCD, then absolute lifetime values were determined by QSSPC, typically in the area of best lifetime in the μPCD image, unless noted otherwise.

μPCD

μPCD stands for "microwave-detected photoconductance decay", which describes the principle how this machine measures effective lifetimes:

A brief ($\approx$ 200 ns) light pulse, typically from a laser in the near infrared (e.g. 904 nm for the system at ISC Konstanz, a Semilab WT-2000, corresponding to an absorption length in Si of 32 μm), illuminates a small area ($\approx$ 1 mm^2) of the wafer and generates excess minority carriers within the sample which cause an increase in the sample's conductivity. The transient decay of this excess photoconductance

2.2. Characterisation of surface passivation layers

is monitored by measuring the change in microwave reflectivity of the sample, using microwave radiation with a frequency tuneable from 10.05 to 10.4 GHz.

Interpretation of the measured decay time constants as effective lifetimes is not straightforward. The microwave reflectance is not linear with the wafer conductance and sensitive to the geometrical arrangement of the sample, microwave antennae and the metallic reflector ("short circuit") behind the wafer [Schoe95]. The sensitivity even changes sign (and so can the decay curve, leading to visible "borderlines" or rings of invalid points on the lifetime map) and must be optimized by adjusting the microwave frequency for different samples.

Due to the non-linearity in microwave reflectance and its dependence on sample thickness and the distance in between sample, the metal back reflector and the microwave waveguide, the resulting photoconductance is not absolute, and thus the injection level is unknown. An iterative procedure to determine the injection level and thus the absolute effective lifetime which then agrees well with the effective lifetime determined by QSSPC has been presented [Ber98,Schmi99], but is not practical for implementation when measuring tens or even hundreds of different samples. Therefore, the μPCD is only used for comparative lifetime measurements in this work, while the absolute effective lifetime is subsequently determined by QSSPC, measuring in the area of the best lifetimes visible in the μPCD lifetime map.

The WT-2000 allows for a spatial resolution of down to 62.5 μm, corresponding to the size of selective emitter or BSF structures. The number of measurement points or pixels in the recorded lifetime image is corresponding to the time it takes to record the image, as the method is serial, i.e. only one spot can be measured at a time and the sample area thus has to be scanned following a line pattern. That means, twice the resolution means about 4 times the measurement time. At a resolution of 2 mm, only 2-3 minutes per sample (156x156 mm^2) are needed, while a resolution of 500 μm requires about half an hour per wafer.

The resolution of choice for a specific measurement is influenced by the nature of the sample and the size of possible structures one wants to observe.

While information in terms of fine structures is naturally lost at coarser resolutions, the average lifetime of the sample is not affected beyond the range of fluctuations observed when repeating a measurement with identical settings. In practice, the differences in between grains in multicrystalline material are well visible with a resolution of 500-1000 μm, and 1-2 mm resolution is sufficient for monocrystalline samples.

QSSPC

The quasi-steady-state photoconductance technique was introduced by R. Sinton [Sin96]. It is now very common in semiconductor and especially photovoltaics research institutions and industry, with more than 600 lifetime-testers currently in use worldwide according to the distributor. This makes the QSSPC well suited

Chapter 2: Surface Passivation and Antireflection Coating

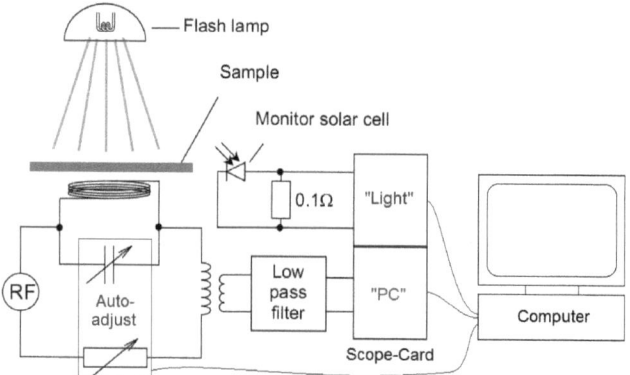

Figure 2.5: *Schematics of the Sinton WCT-120 series QSSPC instrument*

for comparing results with other authors, as it currently is the most commonly quoted source of published effective lifetime values.

The sample is illuminated by a photographic flash with adjustable decay time ($\approx 0.25..12$ ms), situated on top of a coil which is electrically insulated from the sample by a thin plastic layer. The coil is part of a 25 MHz RC bridge oscillator circuit. The resistor and capacitor are automatically balanced by a macro included in the MS Excel-based evaluation software. The oscillating EM-field of the coil couples with free carriers in the sample and thus generates eddy currents which induce a counter-acting current in the coil. After removal of the high frequency part by a low-pass filter, this results in a voltage signal which is almost linearly proportional to the photoconductance (PC) of the sample. Due to the diameter of the coil of ≈ 1.5 cm, the measurement includes a circular area of the sample of $\varnothing \approx 3$ cm. Naturally, the PC value determined in this way is an average value, which leads to errors in case the sample is smaller than the area covered by the coil, or in the case of mc-Si material with several grains of different lifetimes within the measured area.

The light intensity is measured with a gauged reference solar cell of high series resistance which is connected to a fixed load of 0.1 Ω, yielding a voltage signal with a proportionality factor of 5.80 mV/sun for the setup at ISC Konstanz.

Apart from the potential drawback of very low spatial resolution of the measurement, the QSSPC offers the benefits of covering a wide injection range (about 2 orders of magnitude) within a single quick measurement of < 1 s as well as using a light source whose spectrum is at least similar to that of the sun, in contrast to the laser used for the μPCD.

The QSSPC also can be used for resistivity measurements, and here is superior to a 4-point probe setup when it comes to mc-Si material, as the averaging over a larger area levels out some of the doping density differences between single grains

2.2. Characterisation of surface passivation layers

as well as it reduces the disturbing influence of grain boundaries.

2.2.4 Spectroscopic Ellipsometry

Ellipsometry is an optical technique that uses polarized light to characterize thin films. The most important advantage of ellipsometry is its non destructive character and its high sensitivity due to the measurement of the phase of the reflected light, which enables the characterization of very thin layers. The following figure 2.6 shows the principle of a spectroscopic ellipsometer.

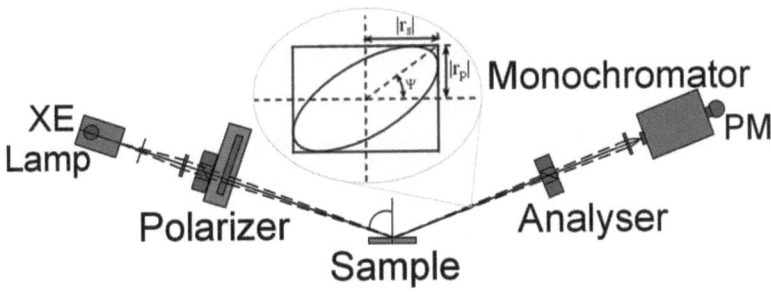

Figure 2.6: *Schematics of a spectroscopic ellipsometer. The sample is illuminated with polarized light, reflected at the layer of interest and finally measured in the detector with an analyzer. From the incident and the reflected light phase change, the layer thickness and refractive indices are calculated.*

In the ellipsometric measurement, incident light is linearly polarized in a polarizer, reflected at the layer of interest and finally measured in the detector with an analyzer. The reflected light has phase changes that are different for electric field components: the p component is oscillating parallel to the plane of incidence and the s component is oscillating perpendicularly to the plane of incidence and parallel to the sample surface. The amplitudes of the p and s components are denoted by r_p and r_s, respectively. Ellipsometry measures the ratio ρ of r_p and r_s

$$\rho = \frac{r_p}{r_s} = \tan\psi \cdot e^{i\Delta} \qquad (2.32)$$

where ψ is the amplitude ratio upon reflection and Δ is the phase change of the p and s components. After measurement, the first step of data analysis is to build a model of the material, including the order of layers, their forecasted thickness and optical constants. After that, the data generated from the model must be compared to the experimental value of $\tan\psi$ and $\cos\Delta$ by a known incident light angle. From this fit, the refractive index and the layer thickness are calculated. This means that the accuracy of the measurements depends not only on the physical properties of the instrument, but also on the correctness of the optical modeling software used for the analysis. It can thus be helpful to apply several models meant to describe the same or similar materials and compare the obtained modeling results in order to get an idea of the possible deviation associated with the different models.

For this work, a spectroscopic ellipsometer Sentech SE800 PV with a xenon lamp as light source (evaluation range $\lambda = 280..930$ nm) was used for characterization of thickness and refractive index of the silicon nitride, oxide and carbide layers.

2.2.5 Fourier Transform Infrared Spectroscopy (FTIR)

Fourier transform infrared (FTIR) spectroscopy is a measurement technique for collecting infrared spectra. Instead of recording the amount of energy absorbed when the frequency of the infrared light is varied (monochromator), the IR light is guided through an interferometer. After passing through the sample, the measured signal is the interferogram. Performing a Fourier transform on this signal data results in a spectrum identical to that from conventional (dispersive) infrared spectroscopy.

FTIR spectrometers are cheaper than conventional spectrometers because building an interferometer is easier than the fabrication of a monochromator. In addition, measurement of a single spectrum is faster for the FTIR technique because the information at all frequencies is collected simultaneously. This allows multiple samples to be collected and averaged together resulting in an improvement in sensitivity.

IR spectroscopy is used to identify chemical compounds based on how infrared radiation is absorbed by the compounds' chemical bonds.

IR absorption in amorphous compound layers of Si with C, N, and O typically takes place in the spectral range of $4000..500$ cm^{-1} (corresponding to $\lambda = 2.5..20$ μm) which was therefore used for the measurements in this work.

3

Fabrication of surface passivation layers

Abstract

This chapter explains the difference between growth and deposition as methods to obtain a film on top of a substrate and gives some examples of the dielectrics and their formation techniques which are most commonly used in current c-Si solar cell technology, with focus on SiO_x and SiN_x from thermal oxidation and Plasma Enhanced Chemical Vapor Deposition (PECVD), respectively.

Subsequently, the thin film formation technology by PECVD is described more detailed, and the particularities of the low-frequency direct-plasma PECVD reactor from Centrotherm, mainly used in this work, are outlined.

3.1 Fabrication of surface coatings-deposition versus growth

Surface coatings can be grown or deposited, applying either solid, liquid or gaseous media as precursors, i.e. source substances. Growing means that the silicon is partially consumed in the growth process by the reaction with another element, e.g. for silicon oxide around 0.45 nm of Si from the surface per 1 nm of grown SiO_2.

While the term "growth" is sometimes also used in a more general way (for deposited layers) in the literature, a clear distinction shall be made within this work.

Depositions are done from one or several precursors, liquid or gaseous, with the precursors containing silicon plus the other elements required for the formation of the desired layers (mostly amorphous compounds with oxygen, nitrogen, carbon and/or fluorine).

The activation energy to start the growth or deposition reactions can be delivered thermally, chemically, radiatively, by ions of sufficient energy in a plasma or a combination of these factors, depending on the environmental conditions.

3.2 Grown films

Films which grow or are grown, respectively, on a silicon surface are mostly either silicon oxide (SiO_2) or, sometimes, stoichiometric silicon nitride (Si_3N_4).

A so-called native oxide film grows naturally on silicon in air even at room temperature, although its thickness is limited to around 2 nm and the quality of this oxide is very low in terms of density and surface passivation. Wet-chemical reactions in acids are another way of growing silicon oxides at room temperature, with potentially much better film quality due to higher density, like e.g. from nitric acid [Mih08].

A third way of oxidizing a silicon surface is by ozone (O_3) treatment, with the ozone created either by electric discharge or ultra-violet (UV) light of sufficiently low wavelengths below 200 nm. The commonly used low-pressure Hg-lamps emit at 185 nm and 254 nm, but only the 185 nm band contributes to excess ozone generation, while the 254 nm part merely splits existing O_3 molecules [Vig76,Tsao07].

But these latter oxides are also limited in their growth, stopping at similar thicknesses as native oxide. The reason is that the growing oxide is of decreasing permeability for further oxygen radicals with increasing thickness, and the energy of these radicals at room temperature in the previously mentioned situations is limited such that the oxide thickness will always remain well below 10 nm.

Growing thicker silicon oxide layers thus requires higher activation energies, which are usually delivered by high temperatures of several hundred degrees Celsius as for dry (in O_2 or N_2O atmosphere) or wet (in H_2O vapor ambient) thermal oxidation. To grow oxides of around 100 nm as is required for single layer antireflection coating (ARC) in a reasonable time for industrial processing (e.g. 30 minutes) requires very high temperatures, >1000°C for dry and still >800°C for wet oxidation. Thermal nitridation, the analogous growth of silicon nitride, requires even higher temperatures of >1100°C, with 1200°C for 4h required to obtain a 50 nm film in nitrogen ambient, which would still be insufficient for antireflection coating [Zhu05]. For lower quality materials like mc-Si and SoG-Si, the high temperatures of thermal oxidation carry the risk of severe deterioration of the bulk lifetime, and the even higher temperatures for thermal nitridation would affect even top quality FZ material.

3.2.1 Thermal oxidation

A standard method to passivate c-Si surfaces is their thermal oxidation at high temperatures (700-1000°C). For many years, thermally grown SiO_2 was the only way to obtain a very good and long-term stable surface passivation of silicon. The best published surface recombination values, obtained on both p- and n-type Si of very high resistivity (>100 Ωcm), are S_{eff} < 10 cm/s[Gruenbaum90, Aberle99]. However, for Si of low resistivities around 1 Ωcm which is the typical range for photovoltaic applications, the passivation quality depends on the c-Si type. This is because the passivation of SiO_2 relies not only on the saturation of dangling bonds, but is also determined by a fixed positive charge near the Si-SiO_2 interface. This charge, which is of an order of magnitude of around 10^{11} cm^{-2}, accumulates electrons at the surface and repels holes. This causes the surface of n-type Si to go into accumulation, which means a very good passivation, independent of the injection level. The surface of p-type Si, on the other hand, is depleted at lower injection levels, causing surface recombination to increase with decreasing excess minority carrier density.

It should be noted that these best passivation values achieved with thermal oxides were obtained by additional annealing in a forming gas ambient (often 5% of hydrogen and 95% nitrogen or argon) at around 400°C.

A relatively simple and effective way to increase the passivation quality of such p-type Si surfaces by thermally grown SiO_2 is the evaporation of a thin (ca. 2 μm) layer of aluminium on top of the oxide, followed by a 20 minute annealing step at 400°C in a nitrogen atmosphere. During this annealing, atomic hydrogen is formed by the oxidation of the aluminium by water molecules previously formed during the SiO_2 growth [Aberle99].

For practical solar cell applications, PECVD-SiN_x deposited on top of a 10 nm thin thermally grown SiO_2 can yield similar H-passivation induced improvements and yield S_{eff} <10 cm/s, or τ_{eff} >1 ms on 4 Ohm.cm p-type Cz-Si after contact firing as was e.g. found at ISC Konstanz (unpublished).

3.2.2 Plasma-activated oxidation

Plasma-activated growth of silicon oxides has been investigated in microelectronics [Kita91] and can offer the advantages of high growth rates at temperatures as low as 100°C with no apparent dependency on substrate temperature [Ske67], but no literature is available on the performance of such layers in solar cells or on their surface passivation quality.

In the framework of this thesis, silicon oxide films of thickness inhomogeneity <1% have been grown in a plasma at 440 degrees from O_2 and N_2O, with and without additional H_2, resulting in average growth rates of about 1.5 nm/min, the same as the best values in [Kita90] and [Kita91] and comparable to wet oxidation at around 860°C or dry oxidation at around 1040°C, depending on the surface doping concentration. Other than with thermal oxidation, the oxide only grows on the wafer surface exposed to the plasma.

However, no surface passivation could be observed by these oxides.

3.3 Deposited films

In contrast to films grown on a substrate, all the component elements for a layer to be deposited are provided from external sources, either right away in the desired final chemical form as in the case of sputtering (e.g. of SiN_x or SiC_x on silicon) or physical vapor deposition (PVD), or contained within liquid or gaseous precursor substances that are brought to react and subsequently form the desired film, as e.g. in chemical vapor deposition (CVD) methods.

3.3.1 CVD-deposition

Chemical vapor deposition (CVD) is a method widely used to grow amorphous films from gaseous precursors. The basic CVD process for producing SiN_x is the reaction of silane (SiH_4) and ammonia (NH_3) at atmospheric pressure and temperatures from 700-1000°C (ACVD-atmospheric pressure chemical vapor deposition).

In low-pressure CVD (LPCVD), silicon nitride films are deposited from dichloro silane (DCS, chemically SiH_2Cl_2) and ammonia at reduced pressure (p$\approx$ 0.1 mbar) and temperatures around 750°C. Lower deposition temperatures around 550-600°C are possible using bis-(tertiary butyl amylo)-silane (BTBAS) instead of DCS [Aru06]. While the higher resulting film density compared to PECVD deposited layers can be beneficial e.g. when using plating for contact formation or improvement, these films generally lack hydrogen and do per se not allow for good electronic surface passivation.

3.3.2 PECVD-deposition

The plasma enhanced CVD (PECVD) reaction of silane, ammonia and optionally nitrogen at a reduced pressure from p≈0.3-3 mbar (≈200-2000 mTorr) allows the deposition of thin silicon nitride films at temperatures well below 500°C. The deposition is possible at room temperature in principle, but these films are of poor mechanical and chemical quality, i.e. not suitable to yield reasonable surface passivation and withstand a contact firing step. The Centrotherm system mostly used for this work gives best results above 400°C, which is just slightly above the temperatures often used in remote plasma systems like from Roth&Rau. These relatively low temperatures are a large advantage for PV technology and semiconductor electronics considering problems associated with high temperature steps such as higher sensitivity to impurities, possible bulk lifetime degradation by thermally induced crystal lattice defects, throughput and profile degradation of a previously diffused emitter and/or front-surface or back-surface field (FSF or BSF).

Due to its advantages, PECVD is now used as a standard tool in semiconductor electronics and PV research facilities and companies around the world. In crystalline silicon PV, it is mainly used for the deposition of amorphous silicon (a-Si:H) and silicon nitride (a-SiN$_x$:H), less often for silicon oxide (a-SiO$_x$:H) and silicon carbide (a-SiC$_x$:H). The latter two are being used so far mainly or, in the case of silicon carbide, only in research facilities. The precursor gases used for the deposition of these films are usually hydrogen containing reactants, resulting in non-stoichiometric films with a content of up to 30% of hydrogen [Cic07].

The process energy required for the chemical reaction of the precursors is delivered by the electromagnetic field that creates the plasma. The plasma, a partially ionized gas, is a mixture of electrons and different plasma radicals with a neutral total plasma charge. Typically, such an artificial plasma is not in thermal equilibrium (i.e. the number of elastic particle collisions with sufficient energy transfer to maintain ionization is not dominating) and electrons and ions recombine continuously. Therefore, the permanent energy input from the electromagnetic field is necessary to maintain the plasma burning. The electrons in a plasma behave similarly as the electron gas in a metal. Plasma is conductive and can interact with and thus absorb electromagnetic waves of frequencies up to the plasma frequency

$$\omega_p = \sqrt{\frac{ne^2}{\epsilon_0 \cdot m_e}} \tag{3.1}$$

where n is the electron density in the plasma, e is the elementary charge and m_e the mass of an electron.

Assuming typical PECVD conditions of p = 1000 mTorr (= 1.3 mbar) and θ = 450°C yields ω_p = 6.82 THz (corresponding to λ = 44.0 µm) for electrons, which lies in the far infrared. Accordingly, the plasma frequency for the relatively heavy ions of silane and ammonia is in the range of 30-40 GHz under the same conditions [Per03].

In contrast, there are cited values of a plasma-frequency of around 4 MHz for a plasma of silane and ammonia as e.g. in [Aberle99], and this is brought up as an advantage of high-frequency (13.56 MHz) PECVD-systems, stating that because of their operating frequency above 4 MHz, they should cause almost no surface bombardment by high-energy ions any more. It could not be clarified by the author how this 4 orders of magitude lower value is determined as no further information is given on the parameters used for the calculation, and assuming the same mass would require charge densities 8 orders of magnitude below the values of the source [Per03].

The plasma contains a lot of energy and can basically be used in two ways:

- "Condensation": direct deposition of particles from the plasma onto a substrate. This means a constant and large supply of reactive particles, but at the cost of the substrate being directly exposed to the plasma, which bears the risk of surface damage induced by high-energy ions.

- "Evaporation": high energy particles are extracted from the plasma and directed towards a substrate outside the plasma (remote-plasma technology). This avoids potential surface damage of the substrate by the plasma, but can cause mixing problems if not all reactants/precursors are directed through the plasma.

3.3.3 Equipment used for this work

The majority of depositions for this work was done in a low-frequency industrial PECVD system from Centrotherm AG, called E2000 HT. This system is used in many industrial cell lines around the world, and thus an interesting tool for research as any improvements developed with this machine could directly be transferred into industrial production. It consists of an electrically resistor-heated quartz tube (l=2000 mm, Ø=300 mm). The direct plasma burns in between the plates of a grahite boat, which is loaded with wafers outside the reactor. Boats of different sizes and thus wafer capacity are available, composed of plates available for various wafer sizes (up to 200x200 mm^2). The following figure 3.1 shows a simplified sketch of the Centrotherm PECVD system.

In this work, two reactors of this type were used, one at the Institute for Physical Electronics (ipe) at the University of Stuttgart, the other at the International Solar Energy Research Center (ISC) Konstanz. They are identical considering tube dimensions and the plasma generator, but the avaibable graphite boats and precursor gases differ. At the ipe Stuttgart, there are two horizontal boats available for round 6 inch (=156 mm)and square-shaped 156x156 mm^2 wafers, each capable of 24 wafers in two columns of 12. At the ISC Konstanz, two identical vertical boats are available for 156x156 mm^2 wafers (capacity 144 wafers each),

3.3. Deposited films

as well as a horizontal boat for 156x156 mm² (capacity 66 wafers) and a vertical boat for 125x125 mm² wafers (capacity 196 wafers).

In general, vertical boats have the advantage of twice the capacity of wafers per given boat dimensions, as both sides of the graphite plates (except for the outmost plates) can hold wafers of the specified size, but smaller wafers cannot be deposited, as they would not completely cover the holes, around 150x150 mm² in size, which have been cut into the graphite plates at the positions intended for the wafers in order to reduce the weight and thermal mass of the boat. A horizontal boat is about twice as heavy as a vertical one of the same size, as the graphite plates may not have holes, for otherwise the electric field above the wafers and thus the deposited film would not be homogeneous. Vertical boats can therefore be heated up and cool down faster (the cooling speed outside the tube is further increased by the chimney effect of the air in between the vertical plates). This reduces waiting times in processing and saves energy. In addition and also associated with the lower thermal mass, it is easier to achieve a homogeneous temperature distribution over the boat, which can also be beneficial for film composition and thickness uniformity, although the effect seems in practice negligible for the horizontal boats and films used and investigated in this work.

Horizontal boats, in contrast to vertical ones, are flexible in wafer size, as any smaller size than the specified one can also be processed on top of a dummy wafer. In this case, the dummy wafer has to be uncoated for optimum conductivity, otherwise the thickness of the film deposited on the top wafer will be lower than desired - and quite likely inhomogenous. Also, the smaller number of wafers (and dummies) is preferable for laboratory size experiments. Consequently, the horizontal boat was mostly used at ISC Konstanz for deposition and growth parameter studies. As it is a drawback of a horizontal boat to be more sensitive to

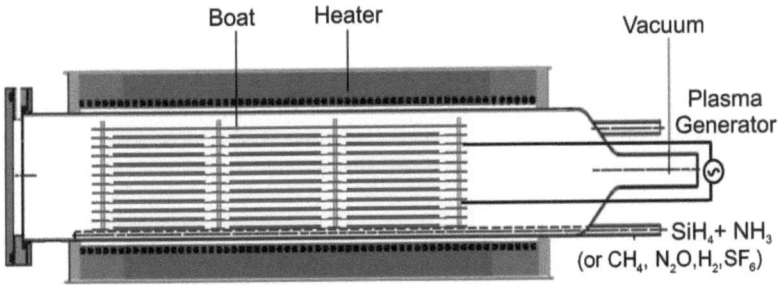

Figure 3.1: *Simplified schematic of the industrial Centrotherm low frequency direct plasma PECVD system used for most experiments in this work. The exemplary horizontal boat in the picture has half the capacity of the actual horizontal boat used for most SiC_x and all SiN_x and SiO_xN_y depositions in this work (after [Cic07]).*

possible gravity-induced gas mixture gradients than a vertical one, the small wafer batches were processed in the same positions of the boat. Results of an exemplary test of homogeneity of the ISC Konstanz standard silicon nitride deposited in the horizontal boat are shown in chapter 4.2

Figure 3.2: *Nitrous oxide plasma burning inside and around a horizontal boat in the Centrotherm PECVD system, as seen through a small quartz glass window in the back end of the tube. The two black lines on the left are the electrical contacts in between the plasma generator and the boat. The light coming from the tube walls is reflected light originating inside the boat. Apparent inhomogeneities in the brightness that may be believed to result from inhomogeneous plasma density in between the plates are actually caused by reflections at the wire mesh acting as RF shielding for the quartz glass window, and by the different viewing angles of the spaces between the different plates. The latter makes a difference as it affects the viewing depth into the plasma which is directly related to the light intensity as light is emitted from everywhere in the mostly transparent plasma volume.*

The vacuum chamber has a volume of ca. 140 liters and contains at standard process conditions (T=450°C, p=224 Pa) $6.18 \cdot 10^{-3}$ number of particles, which results in a particle density of $2.48 \cdot 10^{22}$ $1/m^3$ [Per03]. Under these conditions,

3.3. Deposited films

the mean free path is l = 1.25 mm and the mean thermal velocity v_{th} = 1100 m/s. That yields the mean traveling time of ions in between particle-particle collisions of t = 1.1 µs.

The plasma burns in between the plates of the boats and slightly around the edges and on the outer side of the outmost plates ("edge plasma"). The intensity of the "edge plasma" compared to in between the plates depends on the mean free path and lifetime of the excited ions, and thus the gas mixture and tube pressure. The above figure 3.2 shows the horizontal boat at ISC in a N_2O plasma at a chamber pressure of 1000 mTorr = 1.3 mbar.

The power density of the plasma in the Centrotherm system can be set to values from about 23-230 mW/cm^2 for the large boats at ISC Konstanz, slightly varying in between the different boats, and about 69-690 mW/cm^2 for the small boats at ipe Stuttgart. These values have to be divided by a factor of 11 to give average power densities, however, as the standard recipes for SiN_x depositions in these machines work with a pulsed plasma, i.e. the power is varied with a step-like function, with an on/off-ratio of 1:10.

The purpose of the pulsing is to achieve more homogeneous deposition over the wafer and the boat, as the effect of electric field inhomogeneities is minimized when the field is not present during ca. 90% of the deposition time. In addition, without the electric field present to accelerate ions during the pulse-off time, the possible wafer surface damage by ion-bombardment is greatly reduced.

This may explain why the best surface passivation quality of monocrystalline Si p-type wafers after simulated contact firing achieved by SiN_x and SiO_xN_y deposited with the Centrotherm 40 kHz low-frequency PECVD system within this work (<10 cm/s at both 1 sun illumination and 10^{15} cm^{-3} injection, see sections 4.2 and 5.4) is comparable or equal to that of high-frequency and/or remote plasma systems [Aberle99, Kerr03].

4

PECVD-Silicon Nitride

Abstract

In this chapter, the results of experiments with a-SiN$_x$:H (SiN$_x$) are presented. SiN$_x$ was solely deposited using the low-frequency Centrotherm system, in contrast to SiC$_x$ (Chapter 5).

Regarding deposition parameters, the influence of the gas flow ratio and wafer position in the horizontal boat position on the passivation and optical properties as deposited and after simulated contact co-firing were investigated.

As an approach to potentially reduce the costs of PECVD deposition by using cheaper precursor gases, the effect of a variation of the purity grade of ammonia used for the SiN$_x$ depositions was investigated on lifetime samples as well as solar cells. Finally, the long-term stability of the encapsulated solar cells was tested by temperature variation cycling as no clear difference between the different purity grades was detectable on the cell level between ammonia purity grades N50 (UHP), N36 and N20 (industrial grade, 99% purity).

4.1 Gas flow ratio and substrate quality dependence

The gas flow ratio (GFR) has the largest influence on the film composition in terms of the N/Si-ratio, and thus the refractive index as well as the thermal stability of the electronic surface passivation quality. The GFR defined by (NH_3:SiH_4) was varied from 10 (the standard ratio at ISC) to 1, resulting in films with refractive index $2.03 \leq n \leq 3.3$ (figure 4.1).

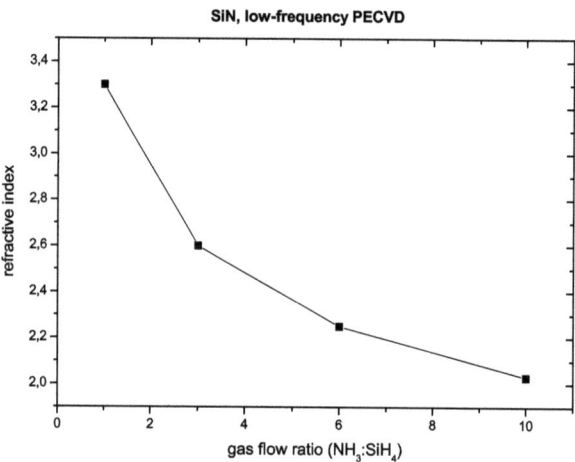

Figure 4.1: *Influence of the gas flow ratio on the refractive index for low-frequency PECVD-SiN deposition.*

Symmetrical lifetime samples were fabricated from two different kinds of FZ wafers:

- NaOH-etched (20 µm/side, resulting in a shiny but not mirror polished surface) 5 inch 2 Ωcm p-type FZ-wafers of 200 µm thickness after etching. The samples were cleaned by HCl (3%, 5 min) and HF (2%, until hydrophobic) prior to SiN-deposition.

- double-side polished 4 inch 2.5 Ωcm 300 µm thick p-type FZ-wafers that were taken directly from the box (i.e. as cleaned by the manufacturer) and only subject to a HF-dip (2%, until hydrophobic) prior to SiN-deposition.

Effective lifetimes were measured as-deposited and after firing (same equipment and parameters as used for metal paste co-firing). The effective lifetimes before and after firing (fig. 4.2) show opposite trends for the 4 and 5 inch FZ-wafers: While the NaOH-etched FZ-wafers all show deterioration of the effective

4.1. Gas flow ratio and substrate quality dependence

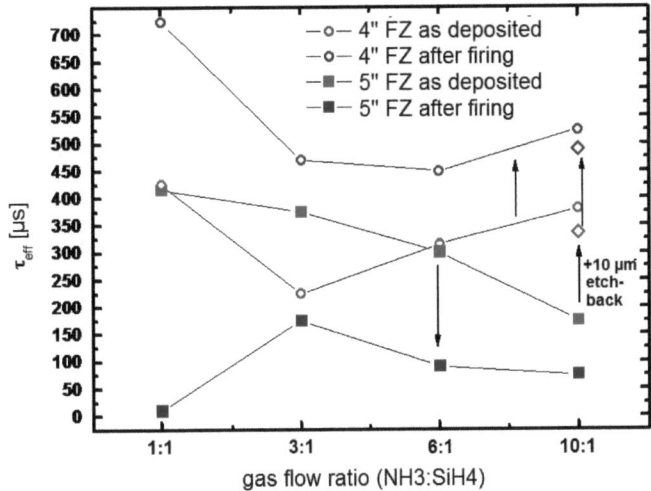

Figure 4.2: *Influence of the gas flow ratio for low-frequency PECVD-SiN deposition on effective lifetimes before and after firing. The negative response to the firing step of the 5 inch samples with 20 µm/side of saw-damage removed as opposed to the positive response of the 4 inch samples is due to a deeply damaged surface of the 5 inch samples. Results with an additional 10 µm/side removed from the 5 inch samples are also shown for SiN (10:1).*

lifetime after firing, all polished FZ-samples benefit from the firing. As both wafer materials are p-type and of similar doping concentration, it seems unlikely that the difference is due to the SiN-films, but rather related to the material or the surface conditions. As the wet-chemical process sequence NaOH+HCl+HF normally allows for improved surface passivation after firing with the standard 10:1 SiN_x at ISC, an influence of the material of the 5 inch wafers seemed to be the most likely reason.

It had been found in previous experiments with lifetime samples that the top 15-20 µm of wire-sawn Cz-wafers sometimes had to be etched back in order to obtain high effective lifetimes. The reason for this is unclear, as the mechanical sawing-induced crystal damage is normally deemed to only affect the first ≈5 µm. A possible explanation is that a wire saw with higher sawing rates was used for these wafers, thus increasing mechanical stress on the wafer surface and inducing deeper micro cracks.

To investigate whether the 5 inch FZ-material had an especially deep "surface"-damage, an additional 10 µm per side were etched back in NaOH from the previously 10:1 SiN-coated sample after SiN_x-removal in HF. Afterwards, the cleaning by HCl+HF was repeated and SiN_x 10:1 was again deposited on both sides.

The resulting effective lifetimes before and after firing of 330 and 480 μs, respectively, are comparable to the observed passivation quality on the 4 inch double-side polished FZ-material. This confirms that the apparent large difference in surface passivation was in fact a material-related effect. Consequently, this 5 inch FZ material was etched back by at least 30 μm per side in future experiments (see chapter 4.4).

Besides the detected influence of near-surface material quality of the substrate, it is also possible that a higher initial surface purity level is required for firing stable SiN$_x$ layers of higher refractive index/lower ammonia to silane gas flow ratio. This can be concluded from the observation that only FZ-wafers taken directly from the box and only dipped in HF gave superior surface passivation both before and after firing for the highest refractive index of 3.3 (GFR=1), while both the HCl+HF cleaned samples and the additionally piranha-cleaned samples of potentially higher surface purity (section 4.2) exhibited the trend of decreasing surface passivation with increasing refractive index. This agrees with the observation that for a-Si, the stoichiometrical upper limit for Si-rich a-SiN$_x$, the surface purity level is highly critical for the success of the subsequent surface passivation [Anger08].

4.2 Boat position dependence

The wafer position within the horizontal boat influences the stoichiometric composition of the deposited film. This is due to gravity de-mixing increased by the horizontal plate configuration and due to a concentration decrease of SiH$_4$ on the way through the plasma volume (resp. the boat) caused by consumption and the lower diffusion and drift velocity of SiH$_4$ due to its higher molecular mass.

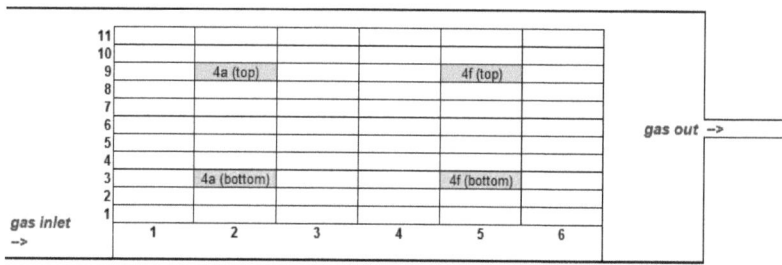

Figure 4.3: *Schematic of the wafer positions used to investigate the homogeneity of the layer properties of the horizontal boat.*

To study the influence of the boat position, four wafers from two different Cz-Si materials (labeled 4a and 4f) with equal resistivity of 5-6 Ω·cm were placed in four different positions of the boat (see figure 4.3) during depositions with

4.2. Boat position dependence

NH$_3$:SiH$_4$ gas flow ratios 10, 5 and 3. Compared to the samples for the previous experiments presented in section 4.1, these wafers received an additional piranha cleaning (10 min, 80°C) and subsequent second HF-dip to reduce the likelyhood of surface impurities potentially damaging surface passivation quality. The samples from material 4a in average showed higher effective lifetimes than samples from material 4f which has to be considered when comparing the boat columns 2 and 5 in terms of surface passivation quality. However, the surfaces were identical and there is no reason to assume an influence of the material on the measured refractive index.

Figure 4.4 and 4.5 show the resulting influence of the boat position on effective lifetimes after firing and refractive index, respectively. The lower boat positions generally give a lower refractive index and higher effective lifetime. The trend of higher effective lifetimes for lower refractive indices both before and after firing is also visible when comparing the three different investigated gas flow ratios.

The change in refractive index after firing is only dependent on the gas flow ratio, except for "gas in, bottom", which is the position close to the gas inlet. This may indicate that gas intermixing of silane and ammonia was not yet complete in this position, however, the effect corresponds to a relative change of only 1% and seems negligible when regarding the total distribution of refractive indices for a given gas flow ratio of 3%, which is also still acceptable for solar cell processing. The variations in thickness (not shown) are below 6%. Both thickness and refractive index inhomogeneity are about 3 times larger compared to the values reported in [Cic07], which makes sense as neighboring columns of a horizontal boat of the same size were investigated in this source, while the columns in this study were 3 columns apart.

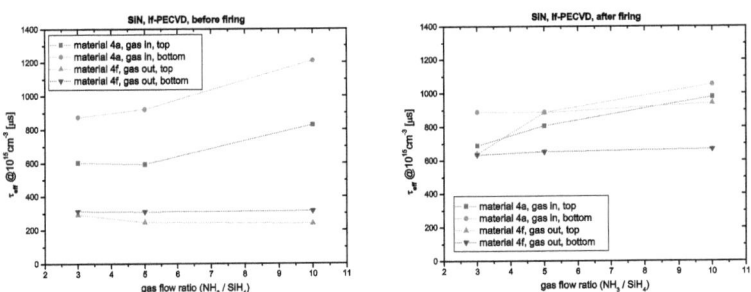

Figure 4.4: *Dependence of the effective lifetime on boat position and gas flow ratio in the horizontal boat. Material 4a in column 2 is near the gas inlet, material 4f in column 5 near the gas outlet (see figure 4.3). These different materials used near the gas inlet and the gas outlet cannot be compared because of their different bulk lifetimes.*

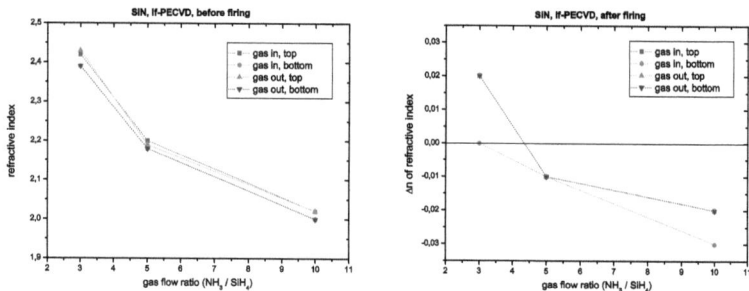

Figure 4.5: *Dependence of the refractive index on boat position and gas flow ratio in the horizontal boat. Relative values compared to values before firing are given for the fired layers to improve visibility. See figure 4.3 for more details on the boat positions.*

While there is no visible decrease in refractive index from gas inlet to gas outlet, the effect of apparently higher silane concentration in the upper part of the boat is counter-intuitive, as one would expect the heavier silane to be found rather in the lower part (the gas inlet is also at the bottom of the tube) and thus higher refractive indices in the lower part of the boat. However, this is consistent with [Cic07] and the results for SiC_xN_y in chapter 5.

A comparison with other published values [Lau97, Kerr03, Cic07] shows that the achieved effective lifetimes on p-type Si exceed previously achieved results for low-frequency PECVD on substrates of comparably low resistivity, and is only slightly lower than the surface passivation quality achieved by high-frequency direct or remote plasma PECVD-deposited SiN [Lau97, Kerr03]. As mentioned in chapter 3.3, this similarity to high-frequency and remote-plasma PECVD may be due to the fact that typically, the electric field is present during less than 10% of the deposition time in average. In turn, this means that remote plasma-like conditions are present over 90% of the time.

4.3 Emitter passivation

Two different $POCl_3$-diffused emitters of initial sheet resistivities of 13 and 63 $\Omega/\square$ on 160 μm thick 6 Ω·cm p-type Cz-Si wafers were etched back by forming and removing porous silicon of different thicknesses. Subsequently, SiN_x was deposited on both sides for lifetime measurements with ISC's standard SiN_x with NH_3:SiH_4 gas flow ratio 10.

Figure 4.6 and 4.7 show the resulting sheet resisitivies and implied V_{oc} values, the average lifetime of the whole wafer measured by μ-PCD, and the lifetime at

4.3. Emitter passivation

10^{15} cm^{-3} injection level on a logarithmic scale. As the implied V_{oc} depends on the natural logarithm of the injection level which in turn is proportional to the lifetime for a given illumination like 1 sun at which the implied V_{oc} is determined, these curves should exhibit linear correlation.

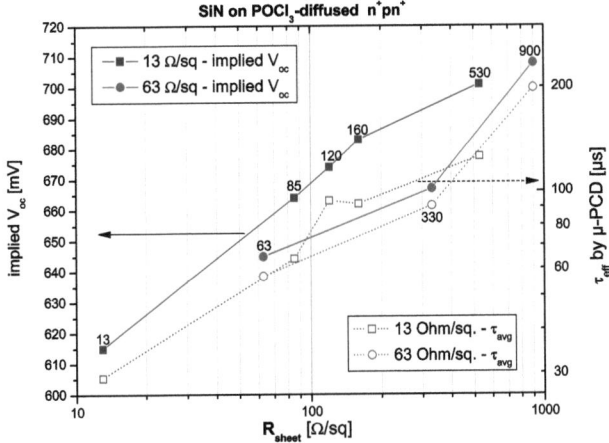

Figure 4.6: *Passivation of various etched-back emitters by PECVD-SiN$_x$, based on a 13 and a 63 $\Omega/\square$ emitter: Implied V_{oc} values of best areas determined by QSSPC compared to average effective lifetimes measured by μPCD*

The most obvious difference between the two curves is the higher effective surface passivation level obtained on samples with the etched back 13 $\Omega/\square$ emitter. This is also visible from the proportionality factor F of the implied V_{oc} to the effective lifetime ($V_{oc} \propto F \cdot \ln \tau_{eff}$) which is higher for the 13 $\Omega/\square$ emitter than for the 63 $\Omega/\square$ emitter.

It can thus be concluded that the bulk lifetime of the wafers was not affected by the higher thermal budget of the stronger diffusion. In this case, the better passivation is to be expected, as an emitter of comparable sheet resistivity etched back from a 13 $\Omega/\square$ will be deeper and have a lower surface doping concentration N_s than if etched back from a 63 $\Omega/\square$ emitter. This is also advantageous for process stability, as the sensitivity of the sheet resistivity to the etching depth is lower.

Regarding the diffusion profiles, only the tail of the P-diffusion profile, created by interstitial diffusion [Ben07], remains for sheet resistivities above 100 $\Omega/\square$. Surface doping concentrations are below $5 \cdot 10^{19}$ and these emitters could therefore not be contacted satisfactory with current screen-printing technology [Stem04], but would require evaporated or plated contacts.

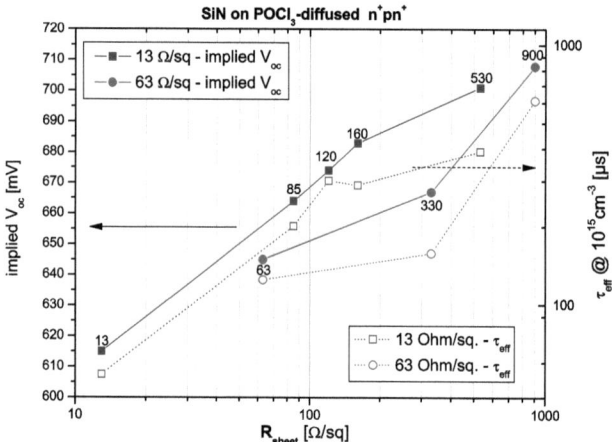

Figure 4.7: *Passivation of various etched-back emitters by PECVD-SiN$_x$, based on a 13 and a 63 $\Omega/\square$ emitter: Implied V_{oc} values of best areas determined by QSSPC compared to effective lifetimes at 10^{15} cm^{-3} injection.*

4.4 Ammonia quality dependence

Usually, the gas quality used for the depositions investigated in this work was ultra-high purity (UHP, or N60 or N50) as it is, to the knowledge of the author, common in laboratories and solar industry around the world to exclude possible effects from other trace elements or substances incorporated into the deposited layers as much as possible. To check whether and up to which point the passivation quality obtained from UHP precursor gases can be maintained before degradation is visible when gradually lowering the gas quality, a study was performed together with Air Liquide, a leading industrial supplier of common precursor gases for SiN$_x$, silane (SiH$_4$) and ammonia (NH$_3$).

In this study, for two reasons, only the quality of ammonia was varied. Firstly, as the plasma for single layer ARC-SiN$_x$ is typically composed of around ten times as much ammonia as silane, ammonia consumption is one order of magnitude larger. Secondly, lower purity grade ammonia is cheaper and could thus reduce production costs because the production process of ammonia (Haber-Bosch synthesis from H$_2$ and N$_2$) is cheaper compared to that of silane (from metallurgical grade Si via trichlorosilane) and the main cost factor is the multi-step purification towards N5.0 or more. Also, due to the synthesis process, there is only one high purity grade of silane fabricated, according to the producer, and thus lower purity grades would have to be manufactured by intentional contamination, which obviously would not make sense, unless a beneficial effect on the resulting layers

4.4. Ammonia quality dependence

could be demonstrated.

To compare various lower purity grades to the standard N5.0 ammonia used at ISC, both solar cells and lifetime samples were fabricated to compare the solar cell characteristics short circuit current density J_{sc}, open circuit voltage V_{oc}, efficiency η and fill factor FF as well as effective lifetimes τ_{eff} before and after a simulated contact firing (same equipment and parameters as used for metal paste co-firing). A possible deterioration of the surface passivation quality should be visible mainly in the V_{oc} of the solar cells and the τ_{eff} of the lifetime samples.

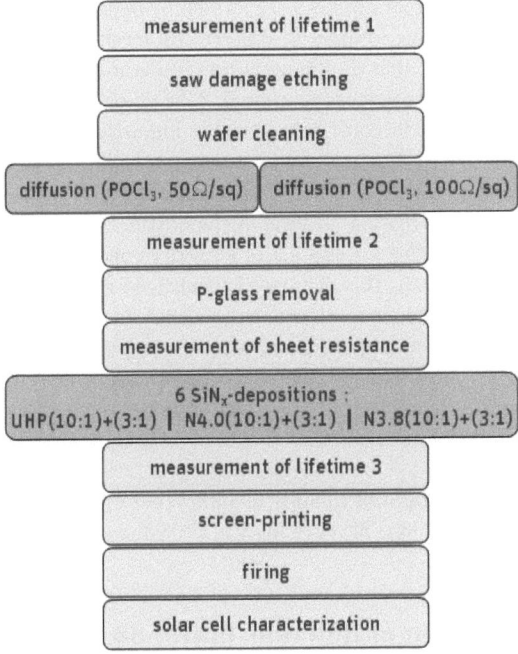

Figure 4.8: *Process flow chart of first series of experiments on ammonia quality: UHP compared to N4.0 and N3.8*

The solar cells featured a 55 $\Omega/\square$ emitter diffusion, which was the ISC standard emitter diffusion at that time. In the first series of experiments, the lifetime samples received a shallow diffusion of 100 $\Omega/\square$ prior to the SiN$_x$-deposition. The reason for using the shallow emitter was to investigate the passivation of the SiN$_x$ on an emitter-diffused surface while being more sensitive to differences in the passivation quality than with the 55 $\Omega/\square$ emitter where the effective lifetime is limited to a lower level by emitter recombination. However, the overall effective lifetimes of the diffused samples were very low and did not allow to see any difference between the groups of different ammonia quality, as the variation in between samples was the same as between single measurements of the same sample. Therefore, the diffused near-surface layer of the lifetime samples was etched back on every second lifetime wafer prior to SiN_x depositions in the second series of experiments, and etched back for all lifetime wafers in the third series.

Figure 4.8 shows the process flow chart of the first series of experiments, where UHP ammonia was compared with purity grades N4.0 and N3.8. No difference was detectable between the different ammonia qualities, neither on lifetime samples nor on solar cell level.

Per consequence, in the seconds series, solely ammonia of quality N36 was investigated, but 4 different groups were formed by depositing the SiN_x films for the solar cells and lifetime samples at different fill levels of the ammonia bottle. The idea behind this is that ammonia is present as a liquid in the pressurized bottle, having a vapor pressure of around 7.3 bar at 15°C (its boiling point at 1013 mbar is at $-33.6°C$ [Nist10]). Possible impurities such as water, higher hydrocarbons and traces of metals should remain preferentially in the liquid phase as they have a lower vapor pressure than ammonia, thus leading to increasing impurity levels with decreasing fill level of the bottle.

According to the tables 4.1 and 4.2 of impurity concentrations for the investigated ammonia cylinders, this behavior is confirmed for most of the impurities which were charted by gas chromatography (except for water which was measured by hygrometry). Consistently, the concentration of N_2, CO and CH_4 decreases with decreasing cylinder fill level due to the higher vapor pressure compared to NH_3. An exception is made by CO_2 which enriches with decreasing fill level, probably due to the higher molecular weight of CO_2 compared to the other substances.

Analyses of trace impurities (mainly metallic) that might be present according to Air Liquide (Li, Na, Mg, Si, P, S, K, Ca, Cr, Mn, Fe, Co, Ni, Cu, Zn) were not carried out.

4.4.1 Comparison of different bottle fill levels for ammonia grade N36

In order to investigate the influence of the bottle fill level for ammonia grade N36, four groups of lifetime and solar cell wafers were coated, using the bottle fill levels listed in the following table 4.3. Lifetime samples were fabricated from FZ-Si wafers, while solar cells were made from mc-Si wafers with resistivities of 1-2 Ωcm.

Again, the ISC standard process was used for the study of the solar cells, and the lifetime sample wafers received a 100 $\Omega/\square$ diffusion. For the lifetime samples, a 70 nm coating layer of 3:1 (NH_3:SiH_4) SiN_x was used instead of 10:1 SiN_x as for the solar cells because this layer deposited with higher silane concentration in the gas mix should contain a higher impurity concentration in the deposited layer compared to the 10:1 SiN_x and thus allow for easier detection of potential

4.4. Ammonia quality dependence

N50 Impurity	Specification [ppm]	Before usage [ppm]	After usage [ppm]
H_2O	<1	<0.2	1.1
H_2			0.1
N_2	<3	<0.2	<0.1
O_2+Ar	<1	<0.2	<0.1
CO	<1	<0.20	<0.05
CO_2	<1	<0.2	0.4
CH_4	<1	<0.05	<0.05

N40 Impurity	Specification [ppm]	Before usage [ppm]	After usage [ppm]
H_2O	<10	3.0	1.1
H_2		0.2	0.4
N_2	<10	12.1	0.2
O_2+Ar	<2	0.2	<0.1
CO	<5	<0.20	<0.05
CO_2	<5	<0.20	0.52
CH_4	<5	2.90	0.09

Table 4.1: *Impurity concentrations within the N5.0 and N4.0 ammonia cylinders before and after the experiments. Note that these values are from single cylinders used for the experiments and concentrations of individual impurities may vary within the specification limits over batches, e.g. be higher or lower in other cylinders.*

effects of impurities on the effective lifetimes.

In contrast to the previous experiments with N50, N40 and N38, the shallow emitter of the lifetime-samples was etched back this time after diffusion and subsequent lifetime measurements, by 1 μm on both samples for SiN D1 and D2, and one sample of D3 and D4, while the second sample of D3 and D4 was etched back by another 10 μm per side. The reason for the changed procedure is that a strong limitation of the effective lifetime of SiN-passivated samples had been observed by even a shallow diffusion like the one used for the lifetime samples in this study in other passivation investigations at ISC in the meantime. Removing the diffused region thus makes the lifetime samples more sensitive to changes in the surface passivation quality by the PECVD-SiN. As the diffusion only influences the dopant density within the first ca. 1 μm from the surface, removing 1 μm should, in principle, avoid its effect if it were only due to the changed dopant

$N36$ Impurity	Specification [ppm]	Before usage [ppm]	After usage [ppm]
H_2O	<1	1.20	20.00
H_2		0.50	2.20
N_2	<3	21.60	0.10
O_2+Ar	<1	<0.1	<0.1
CO	<1	0.20	<0.05
CO_2	<1	0.44	0.64
CH_4	<1	1.10	<0.05

$N20$ Impurity	Specification [ppm]	Before usage [ppm]	After usage [ppm]
H_2O		2.5	7.2
H_2		3.4	0.2
N_2		19.5	0.2
O_2+Ar		1.3	<0.1
CO		<0.05	<0.05
CO_2		0.5	0.34
CH_4		72.89	<0.05

Table 4.2: *Impurity concentrations within the N3.6 and N2.0 ammonia cylinders before and after the experiments. Note that these values are from single cylinders used for the experiments and concentrations of individual impurities may vary within the specification limits over batches, e.g. be higher or lower in other cylinders.*

density. In some previous experiments, however, it was discovered that it can be necessary to remove the top 10-15 μm in order to "see" again the original bulk-lifetime as prior to the diffusion. Obviously, there are additional effects of the diffusion, likely an increase in lattice errors in the regions close to the surface. For this reason, another 15 μm were removed from one of the two samples for depositions D3 and D4 to further increase the sensitivity of the samples after no difference in effective lifetime could be seen in the samples from D1 and D2.

The following figure 4.9 shows the average effective lifetimes measured by μPCD.

While lifetimes are almost identical for the diffused samples, there appears to be a slight increase with decreasing bottle fill-level for the as-deposited SiN_x. After firing, there is almost no difference in between the groups except for the two samples that had an additional 15 μm per side removed by NaOH. The overall

4.4. Ammonia quality dependence

Group	Purity - fill level	Target bottle weight [kg]	Weight before/after depositions [kg]
D1	N36 - 100%	19.00	19.00/18.90
D2	N36 - 50%	15.55	15.45/15.35
D3	N36 - 10%	12.70	12.70/12.50
D4	N36 - 5%	12.35	12.3/12.15

Table 4.3: *Correspondence of group numbers to gas bottle fill levels for ammonia N3.6 in the second round of experiments. The empty bottle weight was 12.05 kg according to the bottle label.*

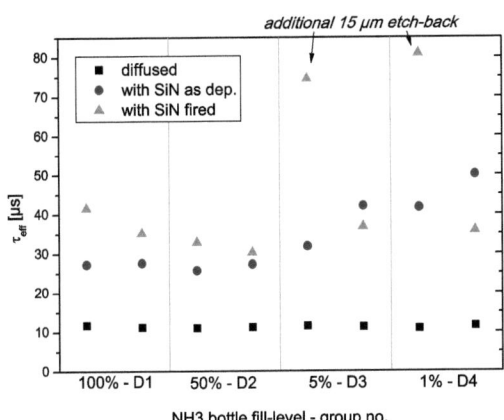

Figure 4.9: *µPCD measured average effective lifetime results for depositions from four different N36 ammonia bottle fill levels. Lifetimes of samples with additional 15 µm etch-back prior to SiN deposition have improved strongly by firing.*

level of these effective lifetimes is very low given that the substrates are FZ Si. As it turned out only after these experiments, the specific FZ material used for these samples had an unusually thick layer of higher recombination activity of more than 25 µm below either surface. The low lifetimes are thus caused by the material, not by the emitter quality and not by the surface passivation quality of the SiN_x. This is also confirmed by the one order of magnitude higher lifetime results in the following third series of ammonia quality experiments where the etch-back thickness of the same FZ material was increased to 50 µm per side for the lifetime samples. Due to the limited number of samples, it cannot be determined whether the observed differences are due to the material quality or the bottle fill-level and thus ammonia purity. An improvement in passivation with decreasing bottle fill-level and thus increasing impurity concentration would contradict the apparent trend in solar cell characteristics in figure 4.10. These were not affected by the problem of insufficient surface damage removal.

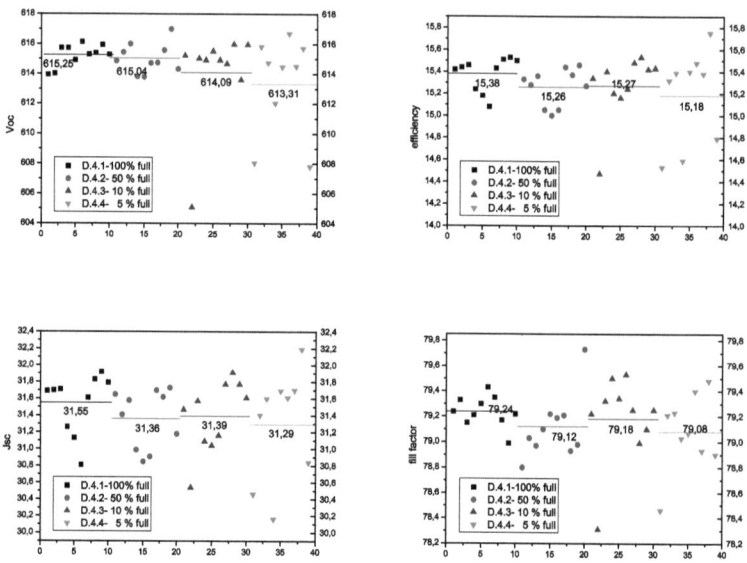

Figure 4.10: *Solar cell results of the second series with different bottle fill levels of ammonia quality N36 on mc-Si wafers. The lines within the groups and the numbers above or below the lines denote the mean value of the respective group.*

Within the solar cells of the second series, there is a tendency of slightly decreasing values of all cell characteristics with decreasing fill level of the N36 ammonia bottle, mainly for the lowest fill levels, in V_{oc} and efficiency. This decrease cannot result from the IV-measurements, as the groups have been mixed prior to measuring in order to avoid any influence of possible drifts of the equipment settings with time. As the two lowest fill level groups have 1 respectively 3 cells with characteristics quite lower than the group's average, it is however possible that the observed tendency would not have been present within a larger group of cells, as the lower fill level groups are, without the outlayers, on the same level as group D1 and D2, and the decrease is also rather marginal. Thus, together with the lack of any deterioration in the lifetime samples from the different groups, it cannot be concluded for sure that there is deterioration in the passivation quality.

4.4.2 Comparison of ammonia grades N50, N36 and N20

As the results of the previous study of N36 ammonia did not clearly answer the question whether there is a degradation in surface passivation with decreasing ammonia quality, a third series of experiments was conducted. A larger batch of samples was prepared for each category this time, and the N36 purity ammonia was compared once more directly to N50 ("UHP1"), and as a novelty also to N20 ("industrial grade"). Five groups were formed, using 225 neighboring mc-Si wafers from two different mc-Si materials and 75 FZ wafers from the same material used for the previous experiments on ammonia purity.

Group	Purity - fill level	Target bottle weight [kg]	Weight before/after depositions [kg]
1	N50 - >50%	n.a.	n.a./n.a.
2	N36 - 50%	15.55	15.6/15.35
3	N20 - 50%	63.8	64.35/64.05
4	N36 - 5%	12.4	12.5/12.25
5	N20 - 5%	44.0	44.1/43.8

Table 4.4: *Correspondence of group numbers to ammonia purity grades and gas bottle fill levels. According to the labels, the weight of the empty N36 bottle was 12.05 kg and the filling added 7 kg; the weight of the empty N20 bottle was 41.8 kg, its filling was 44 kg. The bottle fill level was not determined for N50, but was most likely >50% as this bottle had only recently been installed*

For the depositions of group 1, ammonia of purity N50 was used. Groups 2 and 4 incorporated N36 ammonia, from 50% and 5% fill level of the bottle, respectively, and groups 3 and 5 featured N20 ammonia from 50% and 5% fill level, accordingly. The targeted and actual fill levels before and after the depositions for groups 2-5 are listed in table 4.4. The difference between the target and actual weights is due to the fact that the controlled depletion of the ammonia bottles is time-consuming and that the depletion rate is not linear, but strongly dependent on the temperature of the bottle, and the evaporating liquid ammonia can lead to a significant temperature decrease. Thus, while a maximum depletion rate of 0.35 kg/hour can be reached with a cylinder of 84 liters when starting at 25°C, only depleting for one hour and then waiting for the bottle to reassume ambient temperature, the average depletion rate typically drops to 0.1-0.15 kg/hour after 8h of continuous depletion, depending on the varying ambient temperature during that time. The liquid ammonia within the bottle has then reached about between -25 and -20°C according to the container pressure of about 1.5-2 bars. As the bottle weight can only be determined by uninstalling the ammonia bottle from the gas line and reinstalling it afterwards, it is difficult to reach a precise target weight while using a reasonable number of such installation cycles. Nevertheless, the achieved precision is $<1\%_{rel}$ which seems acceptable and is the same or less than the variation in bottle fill level before and after the depositions.

The mc-Si wafers received acidic iso-texturing. The FZ wafers were etched back by 25 μm per side using 22% NaOH before receiving alkaline random pyramid texturing. This pre-etch was done to ensure good bulk quality also for the solar cells, for which normally only 10μm per side would have been removed by immediate texturing. The reason for this extra step was that this specific FZ-material had exhibited an unusually deep surface damage in previous experiments, such that good effective lifetimes using 3:1 PECVD-SiN$_x$ as surface passivation could only be achieved after etching back said 25 μm per side.

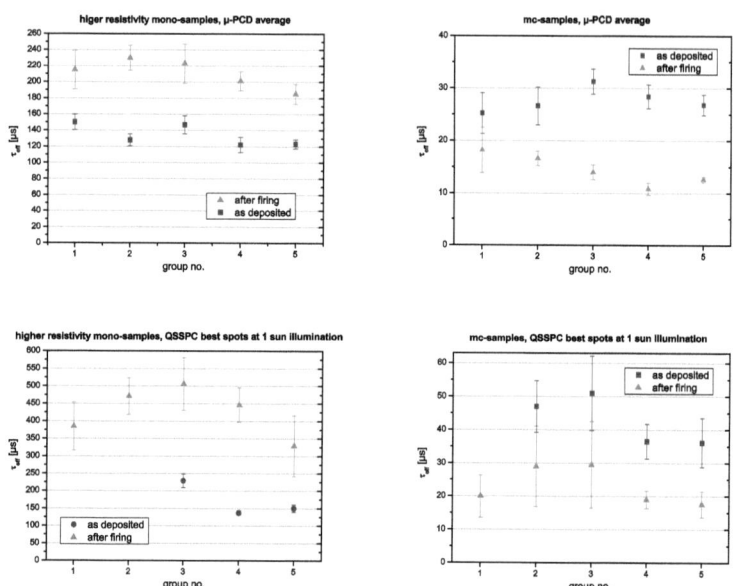

Figure 4.11: *Effective lifetime results of the third series with different ammonia qualities on FZ- and mc-Si wafers. Values are average values and standard deviation from 3 (FZ) and 5 (mc) samples, respectively. The upper two graphs show the average values measured by μPCD, the lower two the values of the best spots from the μPCD image, determined by QSSPC at 1 sun illumination.*

4.4. Ammonia quality dependence

After the texturization, all wafers were diffused in two batches of 150 wafers within two subsequent 55 $\Omega/\square$ standard diffusions, as one diffusion only has space for a maximum of 200 wafers. The wafers were placed into the diffusion boat in their original neighboring order (i.e. as within their original block prior to the wire-sawing) to make sure that neighboring wafers got the most similar diffusion conditions possible. After the diffusion, both the FZ and the mc wafers were split up equally in between the groups by assigning each subsequent wafer to the group of the next higher number, e.g. group 3 after group 2, and starting again at group 1 after group 5.

Five out of the larger group of mc wafers were taken out equidistantly within each group for use as lifetime samples, as well as five of the highest resistivity FZ wafers (five largest code numbers within the batch). These wafers were etched back by another 20 μm per side with NaOH after the phosphorus diffusion to remove the diffused region and the possibly degraded region below it, and the FZ wafers got a final 10 minutes surface cleaning treatment in piranha solution ($H_2SO_4 : H_2O_2 = 4 : 1$, 80°C) prior to the SiN$_x$ depositions. Then, within every group, four PECVD depositions were carried out: 2 for the front and back side of the lifetime samples featuring 20 nm of 3:1 SiN$_x$ followed by 50 nm of 10:1 SiN$_x$, and another two with solely 70 nm of 10:1 SiN for the textured mc- and FZ-Si cell wafers.

As can be seen in the lifetime curves, the full bottles of N36 and N20 give comparable or even better passivation than the N50 reference, while the almost empty bottles yield lower lifetimes. These differences are visible but rather low and given that, in a typical industrial solar cell, the front side passivation has to perform only on top of a diffused emitter with sheet resistances from 40-70 $\Omega/\square$ or surface doping concentrations above 10^{20} cm^{-3}, there is no visible difference to be expected on the solar cell level. Interestingly, the mc samples show lower lifetimes after firing while the FZ samples benefit from the firing step. This can be observed both for the average (μPCD) and the best spot measurements (QSSPC). The reason is not clear, but a material and thus bulk-related effect seems to be the most likely explanation, as both samples received the same surface etching and cleaning treatment steps.

The V_{oc} level of the FZ cells (figure 4.12) and the mc cells out of the two different material (figure 4.13 and 4.14) is very good considering the 55 $\Omega/\square$ emitter and the base material resistivity of 1.5 Ωcm and the variance of below 1 mV between all groups is very low, indicating no visible differences in surface passivation quality. In contrast to what could be expected from the lifetime sample results, groups 4 and 5 have the same or higher average V_{oc} than groups 2 and 3. The constant material quality over the used wafer batch is visible for J_{sc} and V_{oc}.

The variation in the fill factor of the FZ cells (ca. $\pm$ 1%) is twice as high as that of the mc cells ($\pm$ 0.5 %). A possible explanation is that the random pyramid texture of the FZ cells is more sensitive to the detrimental influence

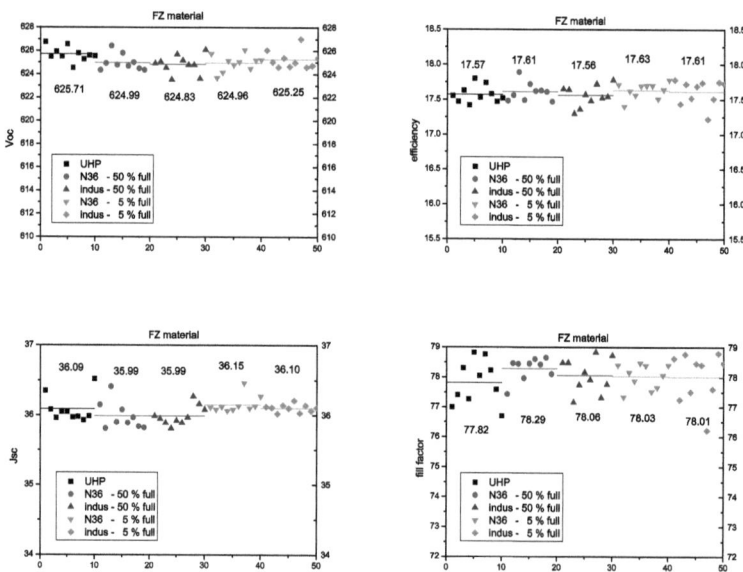

Figure 4.12: *Solar cell results of the third series with different ammonia qualities on FZ-Si wafers. The lines within and the numbers above or below the groups denote the mean value of the respective group.*

of pinholes in the SiN layers, causing higher variations in the contact resistance and/or micro-shunts in areas without SiN coverage.

4.4.3 Module level testing

As no clear detrimental effects of the N36 and N20 ammonia were observed at the cell level, three modules were subsequently fabricated from the FZ- and mc-Si solar cells, and a 200 cycles alternating temperature test was carried out.

Module label	Cell material	No. Cells/string
M1	FZ	8
M2	mc	14
M3	mc	14

Table 4.5: *Modules fabricated for the temperature cycling study.*

4.4. Ammonia quality dependence

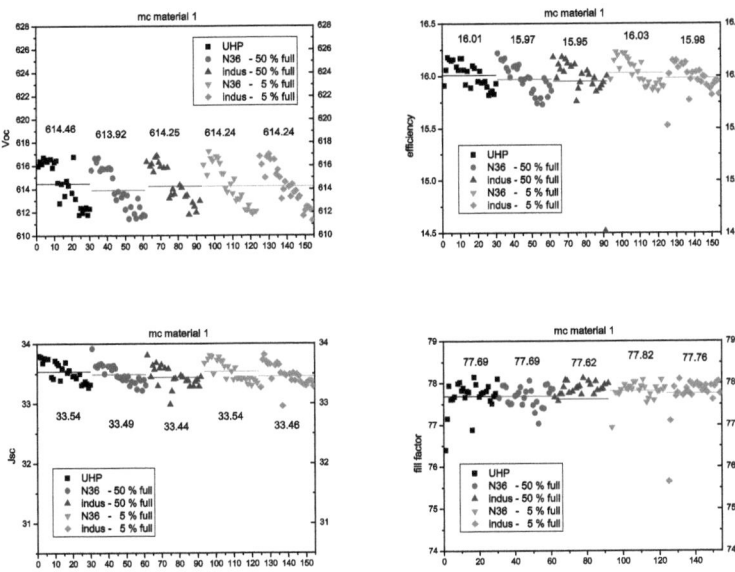

Figure 4.13: *Solar cell results of the third series with different ammonia qualities on mc-Si wafers from material 1. The influence of the ingot height (increasing with wafer number) and thus the material quality is clearly visible for V_{oc}, J_{sc} and efficiency. The lines within and the numbers above or below the groups denote the mean value of the respective group.*

The alternating temperature test is a material fatigue test. The module temperature is altered 200 times from -40°C to 85°C and back to -40°C within 2-7 weeks (depending on the total heat capacity of the investigated modules and the heating and cooling power of the testing equipment), causing strong mechanical stress within the materials of the module. In contrast to UV-degradation testing and damp-heat testing which mainly affect the encapsulation of the module, the main purpose of the alternating temperature test is to investigate whether it inflicts damage on the cells or the other electrical components. To monitor this during the testing, a current is passed through the module and is recorded.

The modules were characterized before and after the temperature cycling by both IV- and electroluminescence measurements to study the influence of the cycling. The tests were carried out in the TestLab PV Modules of the Fraunhofer Institute for Solar Energy Systems (ISE) in Freiburg, Germany, according to the current compliance and approval standard (IEC 61215:2005 10.11b) for terrestrial

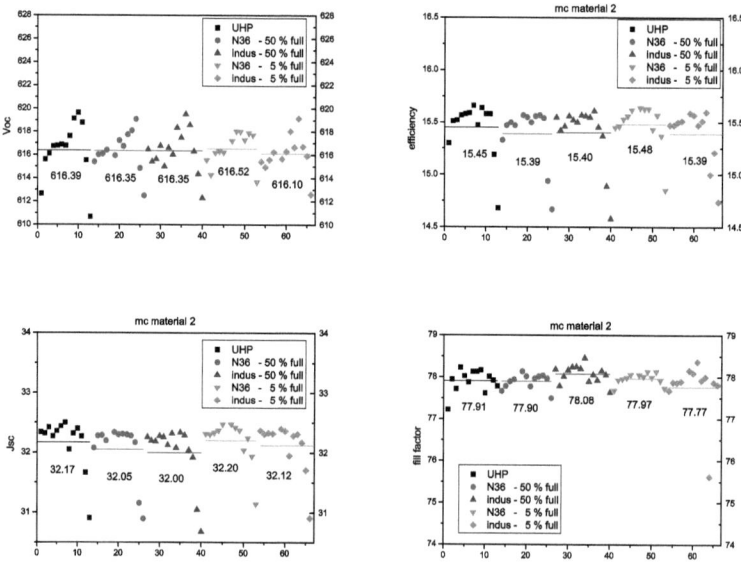

Figure 4.14: *Solar cell results of the third series with different ammonia qualities on mc-Si wafers from material 2. The influence of the ingot height is very similar throughout all groups, but there is no clear tendency as for material 1 (figure 4.13). The lines within and the numbers above or below the groups denote the mean value of the respective group.*

c-Si PV-modules.

Within each module, there were 5 individual strings, each having their own independent connector box (see figure 4.15), with solar cells from one of the five groups each, i.e. solar cells from all five groups were compared within each of the three modules, thus minimizing the potential influence of variations in the lamination process between individual modules. Module M1 was made from 8 FZ-Si cells per strings, modules M2 and M3 from 14 mc-Si cells per string. The modules were fabricated using the proprietary Day4 electrode interconnection technology of Day4Energy Inc. (Burnaby, Canada), thus eliminating the need for Ag-pads on the rear side of the cells and reducing the sensitivity of the strings to soldering problems, as many small wires interconnect the fingers and the fully Al-metallized rear side of the cells in the string instead of only 2 ribbons between busbars and rearside Ag-pads.

4.4. Ammonia quality dependence

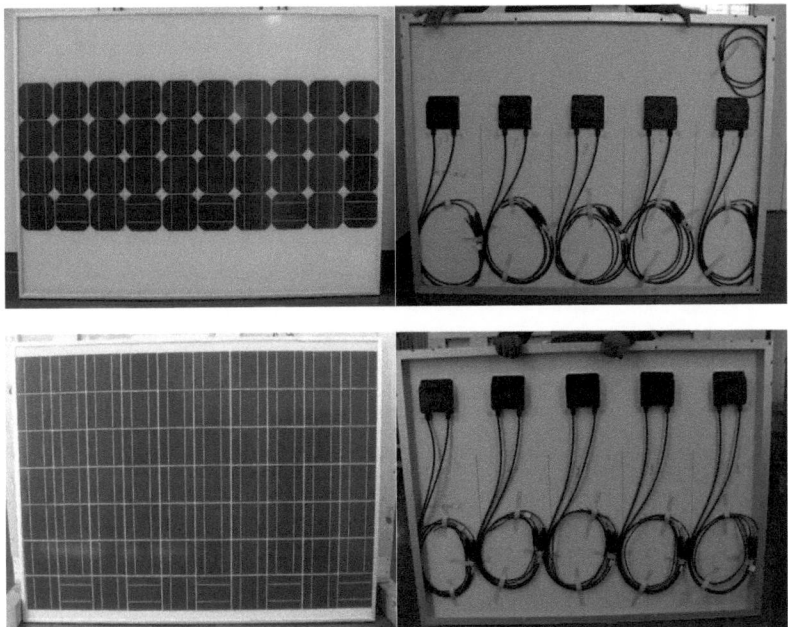

Figure 4.15: *Front and rear view of FZ-cell module M1 (upper images) and mc-cell module M2 (lower images). Strings g1-g5 from left to right on front view images, always containing 2 columns. mc-cell module M3 has identical layout to M2.*

Characterization parameters and results

IV-characterization was carried out at standard testing conditions (AM 1.5 global spectrum, 1000 W/m² illumination, module temperature 25°C).

Electroluminescence (EL) measurements were carried out with a defined current (2 A for M1, 4 A for M2 and M3 due to the larger number of cells per string) being passed through the module (i.e. each individual string in this case) in forward bias, causing near-bandgap IR-emission (i.e. ca. 1150 nm for Si). The light intensity is proportional to the cell current, thus damaged or poorly contacted areas appear darker.

In the upper EL image of FZ module M1 in figure 4.16, a fracture is visible in a cell within string g1. After the 200 cycles of temperature alternation testing, this damage has increased, partially isolating the lower right corner of this cell. In the lower left corner (also string g1), another fracture is visible that almost goes completely across a cell after testing, while it only went across one third of

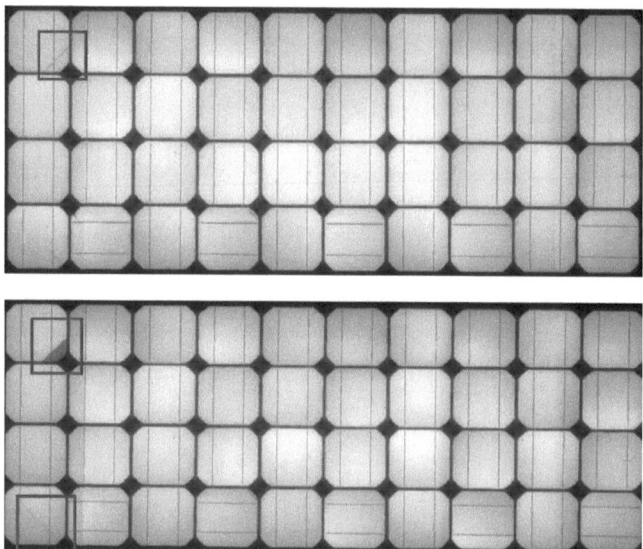

Figure 4.16: *Electroluminescence images of FZ-module M1 before (upper image) and after testing(lower image).*

that cell before. This increase in crack length is to be expected from the high mechanical stress induced by the 200 temperature variations. Despite of these deteriorations, the concerned string g1 shows the second lowest degradation of all strings in this module after testing (see table 4.7).

Module-String	V_{oc}	I_{sc}	V_{mpp}	I_{mpp}	P_{mpp}
M1-g1	-0.60%	0.00%	-2.20%	-0.63%	-2.31%
M1-g2	-0.60%	0.00%	-2.20%	-0.21%	-2.09%
M1-g3	-0.60%	0.20%	-4.39%	0.00%	-4.18%
M1-g4	-0.80%	0.00%	-4.88%	-0.63%	-5.05%
M1-g5	-0.60%	0.00%	-6.10%	-1.06%	-6.58%

Table 4.6: *Relative changes in FZ-module M1 due to temperature cycling.*

In the EL image of mc-module M2 after temperature alternation testing (Fig. 4.17 bottom), all cells appear slightly darker after the 200 cycles (i.e. less current is flowing through them). Apart from this, no difference is visible between the strings that hints an influence of the ammonia quality.

In the EL image of mc-module M3 after temperature alternation testing, all cells appear darker after testing (weaker current flow). Cell 1 from string g3 was

4.4. Ammonia quality dependence

Module-String	V_{oc}	I_{sc}	V_{mpp}	I_{mpp}	P_{mpp}
M2-g1	-0.47%	-0.39%	-2.68%	-0.21%	-2.43%
M2-g2	-0.47%	-0.20%	-1.83%	-0.42%	-1.89%
M2-g3	-0.47%	0.00%	-3.52%	-0.43%	-3.25%
M2-g4	-0.47%	-0.40%	-2.68%	-0.43%	-2.74%
M2-g5	-0.47%	-0.40%	-4.23%	-0.64%	-4.27%

Table 4.7: *Relative changes in mc-module M2 due to temperature cycling.*

Module-String	V_{oc}	I_{sc}	V_{mpp}	I_{mpp}	P_{mpp}
M3-g1	0.24%	-2.24%	-0.99%	-0.87%	-1.88%
M3-g2	0.24%	-1.65%	-2.39%	-1.10%	-3.25%
M3-g3	(6.38%)	(-2.13%)	(4.85%)	(-2.68%)	(2.34%)
M3-g4	0.24%	-1.23%	-1.97%	-0.88%	-2.69%
M3-g5	0.24%	-1.64%	-1.41%	-0.66%	-2.08%

Table 4.8: *Relative changes in mc-module M3 due to temperature cycling. Note that M3-g3 was influenced by one cell from the string only being electrically active after temperature cycling, thus the results of this string cannot be compared with the others.*

without current (not contacted) before the testing. After the 200 cycles, the cell was contacted and showed current flow for unknown reasons, as soldering should not have occurred at the presumed maximum of 85°C during testing, although the exact contacting process of the Day4electrodes has not been published as it is proprietary. The contribution of the additional cell can also be seen in the power measurement and thus this string is not comparable to the others, as the influence of the previously non-contacted cell is unknown.

4.4.4 Conclusions

All modules show power degradations after the temperature cycling. The performance losses are mainly due to decreasing V_{mpp}, not V_{oc} values, which means an increase in series resistance instead of a decrease in surface passivation quality. This can be caused either by changes in the cell metallization or the cell interconnection.

Even though no solar cells were made from the almost empty N50 bottle within this study, based on N36 and industrial grade ammonia results, it should be noted that cells made with ammonia supply from almost empty cylinders

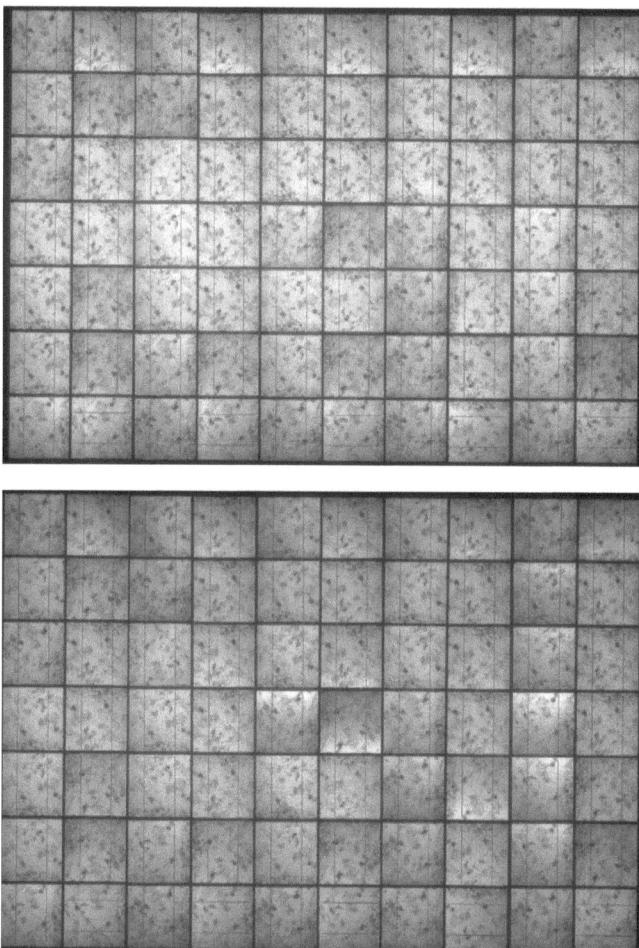

Figure 4.17: *Electroluminescence images of mc-module M2 before (upper image) and after temperature cycling(lower image).*

gave strings with lower performance for modules M1 and M2. However, the effect was not observable for module M3 which should be comparable to M2, and the performance loss is mainly due to a series resistance increase (lower V_{mpp}) while the V_{oc} changes seem to be unrelated to the ammonia quality. Thus, the conclusion is that there can be an impact on the cell-interconnection in some cases, but not on surface passivation. Further investigations should be carried out, preferably with larger solar cell batches and also comparing at least two different module interconnection methods, one of which does not make contact

4.4. Ammonia quality dependence

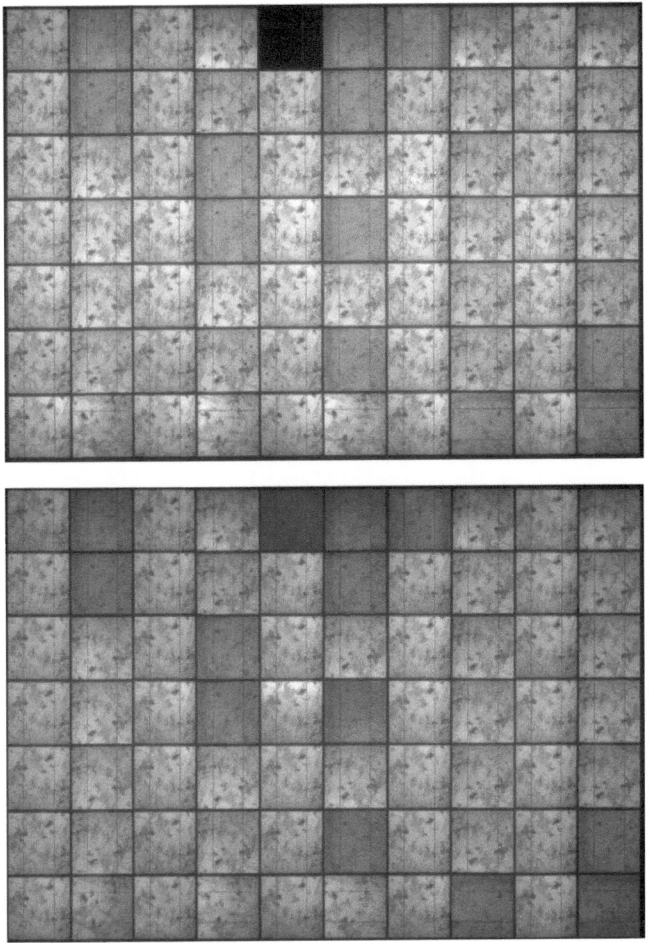

Figure 4.18: *Electroluminescence images of mc-module M3 before (upper image) and after temperature cycling(lower image).*

to the SiN_x layer as does the method used for the modules M1-M3, to check whether this actually has an influence.

This work demonstrates that N50 NH_3 is not mandatory to produce good performance c-Si solar cells and modules. From the present study, based on long-term performance characterization of the modules assembled from cells manufactured with various ammonia grades, it seems reasonable to consider usage of N36 or higher grades of NH_3 for industrial manufacturing of SiN layers. From the industrial perspective, the recommendations prior to further investigations would be to avoid industrial grade ammonia and prevent any ammonia dry-out because that could result not only in a temporary line shut-down, but also in possible off-specs solar cells.

5

PECVD-Silicon Carbide and Silicon Carbonitride

Abstract

This chapter presents the experiments and results of a-SiC$_x$:H (SiC$_x$) depositions from methane and silane by both 13.56 MHz high-frequency and 40 kHz low-frequency PECVD, as well as experiments with a-SiC$_x$N$_y$:H (SiC$_x$N$_y$) deposited by low-frequency PECVD from either methane, ammonia and silane or ammonia together with an alternative silicon-containing precursor.

Resulting films were characterized by QSSPC lifetime measurements and thickness and refractive index evaluated by spectroscopic ellipsometry. The low-frequency deposited samples were later also subject to FTIR measurements to investigate their structure and chemical composition.

The best passivation of τ_{eff}=4 ms at 10^{15} cm^{-3} / S$_{eff}$ <4 cm/s for as-deposited SiC$_x$ on n-type FZ-Si (p-type: τ_{eff}=2.7 ms at 10^{15} cm^{-3} / S$_{eff}$ <5 cm/s) was found for layers fabricated by the high-frequency PECVD with highest Si content (methane/silane gas flow ratio GFR=1), i.e. these layers were stoichiometrically close to amorphous silicon, with the passivation quality continuously decreasing with Si content. This trend changed after firing, with the highest effective lifetime on n-type FZ-Si of 142 μs (S$_{eff}$ <100 cm/s) being reached at a GFR of 6, whereas the passivation of the 4 ms sample almost completely degraded to 4 μs, which is in accordance with a-Si:H results, not able to withstand firing.

SiC$_x$ from low-frequency PECVD gave lower passivation for the as-deposited layers with a maximum of τ_{eff}=400 μs at 10^{15} cm^{-3} / S$_{eff}$ <20 cm/s on p-type Cz-Si, but the best passivation after firing of τ_{eff}=220 μs at 10^{15} cm^{-3} - S$_{eff}$ <30 cm/s is better than for the films deposited by high-frequency PECVD within this work, and equals the best passivation by intrinsic SiC$_x$ reported in [Martin03] and [Ferre08].

5.1 PECVD-SiC$_x$ for surface passivation

Early work on PECVD-SiC$_x$ for c-Si surface passivation using a 13.56 MHz parallel plate laboratory PECVD system was carried out by M. Vetter at the ipe Stuttgart in the late 1990s. He continued it with fellow researchers at the Universitat Politecnica de Catalunya (UPC) in Barcelona. Various publications [Martin01, Martin02, Martin03] demonstrated excellent surface passivation by SiC$_x$ films for both p- and n-type FZ wafers. As the reported surface charge at the Si/SiC$_x$-interface of around 10^{11} cm^{-2} is one order of magnitude lower than for PECVD-SiN$_x$ [Martin03], PECVD-SiC$_x$ was subsequently also investigated on p$^+$ emitters by the author [Petres05a, Petres05b, Petres06], as it was believed that the higher fixed positive surface charge of PECVD-SiN$_x$ is responsible for the non-existent passivation or even de-passivating effect of PECVD-SiN$-x$ on p$^+$ [Kerr03, Libal05, Petres05a].

5.2 High Frequency Direct Plasma

Two different industrial high-frequency (13.56 MHz) direct-plasma PECVD reactors were used for experiments.

5.2.1 p$^+$-Si Passivation

The passivation quality achieved with SiC$_x$ on p$^+$ [Petres06] using a 13.56 MHz parallel-plate prototype reactor from Centrotherm was better than that of the SiN$_x$ films used for comparison, but still was moderate and not sufficient for good solar cells. The following figure 5.1 shows the implied V$_{oc}$ values achieved by stacks of intrinsic and boron-doped SiC$_x$ on symmetrical p$^+$np$^+$ lifetime samples, compared to SiN$_x$ and a 10 nm thin dry-thermal SiO$_2$ as well as the "unpassivated" state, i.e. with the boron-silicate glass (BSG) removed by HF and allowing >1 day for formation of a native oxide layer. Stacks of about 10 nm of Si-rich SiC$_x$ (n$\approx$2.6) with a top layer of 60 nm C-rich SiC$_x$ (n$\approx$2.0) were used for SiC$_x$ deposition as it had been found that while the Si-rich layer yielded good surface passivation before firing, short-wavelength absorption was high down from 400 nm, while the C-rich layer, although not providing any surface passivation, was transparent down to 300 nm. The samples were made of 2.8 Ω·cm n-type Cz with a 90 $\Omega/\square$ BBr$_3$-diffused emitter with a surface doping concentration of $5 \cdot 10^{19}$cm^{-3}.

5.2.2 p- and n-type Si passivation

A 13.56 MHz PECVD system from Applied Materials was used for a systematic study of the surface passivation of SiC$_x$ on p- and n-type FZ-wafers.

5.2. High Frequency Direct Plasma

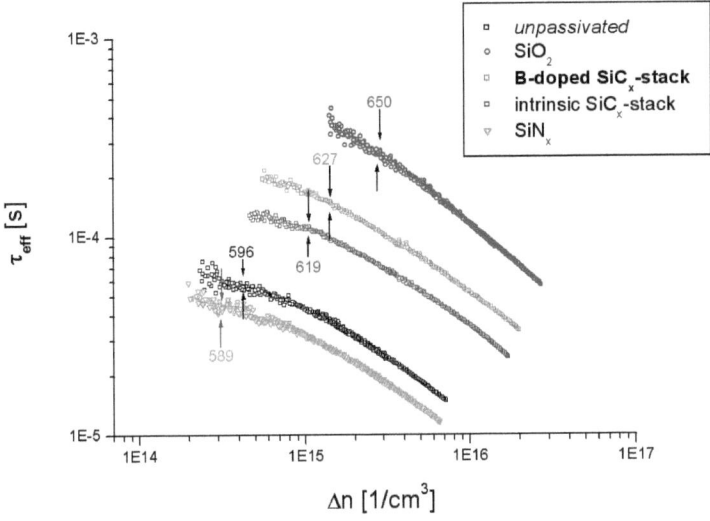

Figure 5.1: *Surface passivation of a symmetrically BBr$_3$-diffused n-type Cz-wafer by different dielectrics compared to the bare emitter with native oxide ("unpassivated"). SiC$_x$ films were deposited by hf-PECVD, SiN$_x$ was deposited with an industrial lf-PECVD that normally yields very good surface passivation of p-, n- and n$^+$-Si (see chapter 5.3). Values are for as-deposited layers. Arrows indicate the injection level at 1 sun illumination by the QSSPC's flash, numbers at arrows denote the corresponding implied V$_{oc}$*

Sample preparation

For sample preparation, <100> oriented n- and p-type FZ wafers were used. Wafer resistivity was 1.5 Ωcm for n-type and 2.5 Ωcm for p-type material which corresponds to a doping density of about $5 \cdot 10^{15}$cm^{-3}]. Wafer thickness was 260 to 280 μm for n-type and 275 to 325 μm for p-type. The samples were cleaned using hot (80°C) piranha (H$_2$O$_2$ + H$_2$SO$_4$ (1:4)). The resulting thin wet-chemically grown oxide was left on the wafers during transfer to Applied Materials for surface protection. Directly before introducing the wafers into the PECVD reactor, the wafers were etched in HF (5%) to remove the oxide and guarantee a clean surface.

SiC layer deposition

Silicon carbide layers of about 100 nm thickness were deposited on both sides of the wafers at Applied Materials. As variations in gas flow ratio, power and pressure change the deposition rate, deposition time was adjusted to yield the desired thickness after an extra test deposition with the targeted parameters carried out beforehand. Depositions were made with different methane to silane gas flow ratios as well as values for radio frequency power and chamber pressure. Table 5.1 contains the studied parameters and their range of variation.

Factors	Range	Levels
Gas flow ratio GFR(CH_4/SiH_4)	1 - 21.2	6
RF power [%]	100 - 400	4
Pressure p [%]	100 - 500	5

Table 5.1: *Parameters used for SiC_x depositions in the Applied Materials high frequency PECVD system*

Lifetime measurements were carried out before and after firing (same equipment and parameters as used for metal paste co-firing) using the Quasi Steady State Photoconductance technique (QSSPC). Thickness, refractive index at $\lambda = 600$ nm as well as the extinction coefficient k at $\lambda = 400$ nm were determined by spectroscopic ellipsometry by Applied Materials.

Influence of Plasma Power

The first series of depositions focused on the influence of the plasma power, starting with the highest methane/silane gas flow ratio (GFR) possible with this PECVD system, 21.2.

The overall level of surface passivation achieved within this first series of experiments is low, and all films exhibit degradation of the surface passivation by the firing step, approaching the level without surface passivation. The dependency is inverted by the firing step: While an increase in plasma power leads to a decrease in passivation for the as-deposited films, the fired layers show higher lifetimes with increasing power.

Plasma power has a clear influence on the optical properties of the SiC_x films. Both the refractive index n at $\lambda = 600$ nm and the extinction coefficient k at $\lambda = 400$ nm decrease linearly with increasing power. While the visible but minor total decrease in refractive index of only -0.03 when quadrupling the power hints at a comparably minor increase in carbon concentration and thus a small change in the stoichiometry of the film, the extinction is reduced remarkably by $30\%_{rel}$. This corresponds to the decrease in extinction coefficient obtained by increasing the pressure mentioned in the following section, but without the associated strong decrease in refractive index. Presumably, this points at differences in the density and molecular configuration of the layer. In conclusion

- The power is influencing the structure of the film rather than its stoichiometry

5.2. High Frequency Direct Plasma

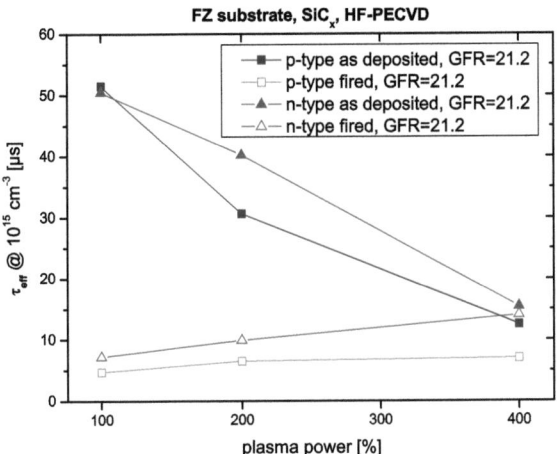

Figure 5.2: *Effective lifetime results of the first series of SiC experiments with high-frequency PECVD on FZ wafers, varying the plasma power. Values are from single wafers.*

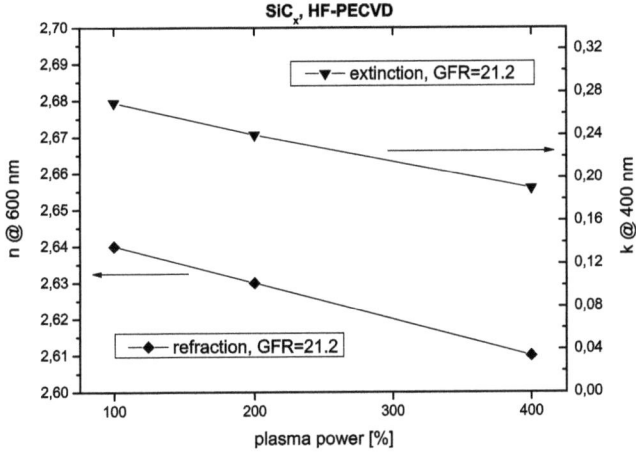

Figure 5.3: *Influence of the plasma power on the optical properties of a-SiC_x:H deposited by high-frequency PECVD.*

- the film structure has a higher influence on the surface passivation quality than the stoichiometry.

Pressure

The chamber pressure was subsequently varied from 100% to 500% of the standard value. Figures 5.4 and 5.5 show the resulting influence on the passivation and optical properties.

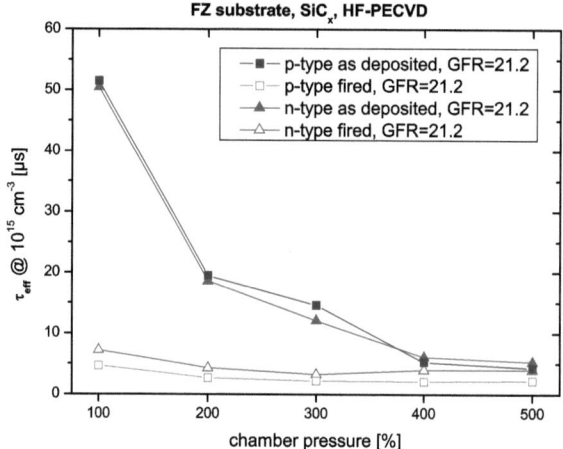

Figure 5.4: *Influence of the chamber pressure on the passivation properties of a-SiC_x:H deposited by high-frequency PECVD on n- and p-type FZ wafers. Values are from single wafers.*

Similar to an increase in plasma power, increasing chamber pressure causes a decrease in surface passivation up to the point where surface passivation both as-deposited and after firing is almost non-existent. Again, all films degrade by firing, but those deposited on p-type wafers to a lower level than those on n-type wafers. Assuming the same S_{eff} at the surface of the samples, this higher lifetime is to be expected from the lower minority carrier diffusion constant in n-type, which is not completely compensated by the lower thickness of the n-type wafers used in the experiments.

An interesting difference between the influences of plasma power and chamber pressure is the fact that while the observed decrease in surface passivation is similar, and the decrease in the extinction coefficient is almost identical, the refractive index decreased much more with increasing chamber pressure, from 2.64 to 2.18, while with increasing plasma power, it only decreased from 2.64 to 2.61.

5.2. High Frequency Direct Plasma

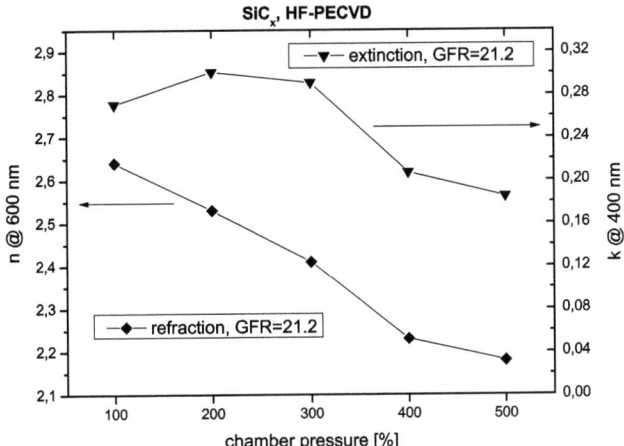

Figure 5.5: *Influence of the chamber pressure on the optical properties of SiC_x deposited by high-frequency PECVD on n- and p-type FZ wafers. Values are from single wafers.*

This points to the film stoichiometry being more closely linked to the refractive index than the extinction coefficient. The extinction coefficient seems to be more influenced by the film microstructure that is related to the surface passivation quality. This is also shown in [Ehling10], where a lower film density caused by e.g. higher plasma power density or C-content was found to be detrimental for surface passivation quality.

Gas flow ratio

The CH_4:SiH_4 gas flow ratio (GFR) was the parameter investigated in the widest range and at the most levels, as it was expected to have the largest influence on film stoichiometry and thus the optical and likely also the passivation properties.

The GFR has a strong influence on the film properties, with its tendency being strongly affected by the firing step.

While the passivation quality of the as-deposited layers steadily increases with decreasing GFR and thus increasing Si concentration in the film, it peaks for the fired layers at a GFR of 6 and is almost non-existent for the extremal GFR values of 1 and 21.2. The passivation of all films effectively degrades due to the firing step, the best passivation achieved at a GFR of 6 is equivalent to a SRV of above 100 cm/s, which is not sufficient for an effective rear surface passivation layer without an underlying BSF diffusion.

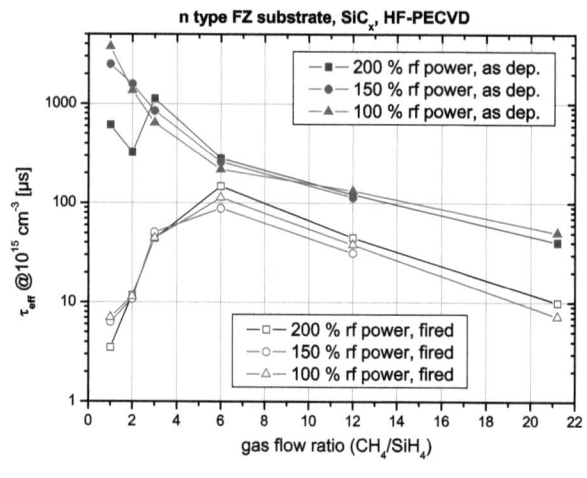

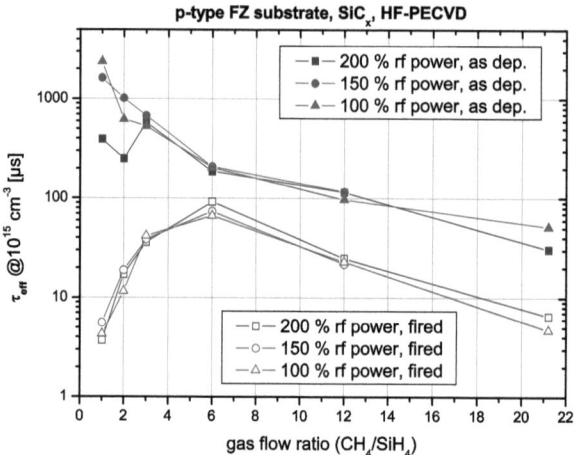

Figure 5.6: *Influence of the gas flow ratio on the passivation properties of SiC_x deposited by high-frequency PECVD on n- and p-type FZ wafers. Values are from single wafers.*

Interestingly, the effect of deposition power is minor in this context, and it appears that the films deposited at the highest power density, while showing the lowest surface passivation as-deposited, are the most firing stable ones and show highest lifetimes after firing on both p- and n-type surfaces. This tendency is consistent with the influence of increasing plasma power observed in the first series of experiments, using a GFR of 21.2.

5.2.3 Etching behavior

It should also be mentioned that PECVD-SiC$_x$ films, unless very rich in Si, are, according to literature, resistant to any wet-chemical etching solution that is currently used in c-Si solar cell processing and would require extreme temperatures, electrochemical or plasma etching approaches [Janz06]. The author got consistent results by placing Si wafers that were completely coated by C-rich SiC$_x$ deposited with GFR=21.2 in HF (5%), NaOH (20%) and CP6 for more than 48 hours, without observing a detectable etching effect. This can be advantageous if an effective etch barrier is desired, but a problem when it comes to cleaning the processing chamber of the PECVD reactor.

5.3 Low Frequency Direct Plasma

5.3.1 First experiments

For the deposition of SiC$_x$ films by the low-frequency Centrotherm PECVD system (see section 3.3.3), the chamber pressure, substrate temperature, plasma power and CH$_4$:SiH$_4$ gas flow ratio were varied, as they have been found to be significant for HF-PECVD SiC$_x$-layers passivation layers in the previous experiments in this work as well as [Martin03, Janz06, Laveuve07, Ferre08]. Note that parameters yielding good passivation on the high-frequency systems could not be tested on this low-frequency system, because some of the settings, e.g. a pressure of 300-500 mTorr, would not allow for a homogeneously burning plasma all over the boat in the low-frequency system.

Factors	Range	Levels
SiC$_x$ - Gas flow ratio (CH$_4$/SiH$_4$)	4 - 28	3
SiC$_x$N$_y$ - Gas flow ratio (CH$_4$/NH$_3$)	0.5 - 4	3
RF power P[W]	2800	1
Pressure p[mTorr]	1000 - 1600	2
Temperature T[°C]	350 - 450	3
relative RF power on-time [%]	100 - 500	3

Table 5.2: *Parameters used in the first series of SiC$_x$ depositions in the Centrotherm low-frequency PECVD system at ipe Stuttgart*

All depositions were carried out on double side polished (DSP) 4" p-type

FZ-wafers. The wafers were taken directly from the delivery box and immersed into fresh 5% HF for a 15 s dip prior to being loaded into the horizontal boat. Two wafers were used per set of deposition parameters, one of which rested on a 156x156 mm^2 p-type Si wafer as substrate, the other on a stainless steel support ("Fe-carrier"). In contrast to when using a Si wafer as substrate, the FZ wafer situated on the Fe-carrier carrier only makes contact to the substrate on its outmost edge, in the shape of a 2 mm wide ring. This should minimize the possibility of contamination of the rear side of the sample which would normally make contact to the substrate wafer. Thus, when using the Fe-carrier, the conditions should be more comparable to using a vertical boat (which is more common in industrial production as mentioned before), where contact to the graphite plate (slightly SiN_x coated) is also made only by the outer edge of the wafer. The alternative approach of using a horizontal boat with special plates made to fit round 4" wafers and thus making contact only through three pins near the outer edge was not available for financial reasons.

The SiC_x deposition recipe started with an ammonia plasma pre-treatment identical to the one normally used for SiN_x depositions. The purpose of such a plasma pretreatment is to slightly etch back and thus clean the wafer surface. Subsequently, ammonia was switched off and silane and methane were introduced to begin the deposition. Due to the chamber volume and gas flows, the content of the chamber was replaced within less than 100 ms (dwell time of a molecule in the tube is typically less than 100 ms), meaning that not even a monolayer (ca. 0.2 nm) could have been deposited from silane together with the remaining ammonia in the switch-over time, as the deposition rate was typically around 0.1 nm/sec.

In parallel to SiC_x, SiC_xN_y was also investigated. It was deposited using silane, methane and ammonia as precursors. The parameters were varied similar to SiC_x, varying the gas flow ratio of methane to ammonia instead of the ratio of methane to silane (see again table 5.2).

The deposition time was fixed to 600 s for the first side, targeting a thickness of 70-90 nmm and adapted for the backside in case the deposited layer was notably thicker or thinner than 70-90 nm, as determined from its color and the intensity of this color as no ellipsometer was available at that time*.

Anyway, no change in the passivation quality was expected for film thicknesses above 40 nm, based on the results of [Martin03, Kerr03, Ferre08] with PECVD-deposited SiN_x, SiC_xN_y and SiC_x.

---

*Practical experience with different samples with thicknesses of 30-130 nm and refractive index of 1.6-3.2 has shown the precision of the experienced human eye for this purpose, and it has been found that it is possible to determine the thickness with an error Δd of about ± 10 nm and the refractive index with an error Δn of about ± 0.5. The precision was highest around the optimum ARC conditions for Si, d=75 nm and n=1.95 (determined at λ=600 nm) with Δd of about 5 nm, which is due to the larger color variation rate with thickness in this regime.

5.3. Low Frequency Direct Plasma

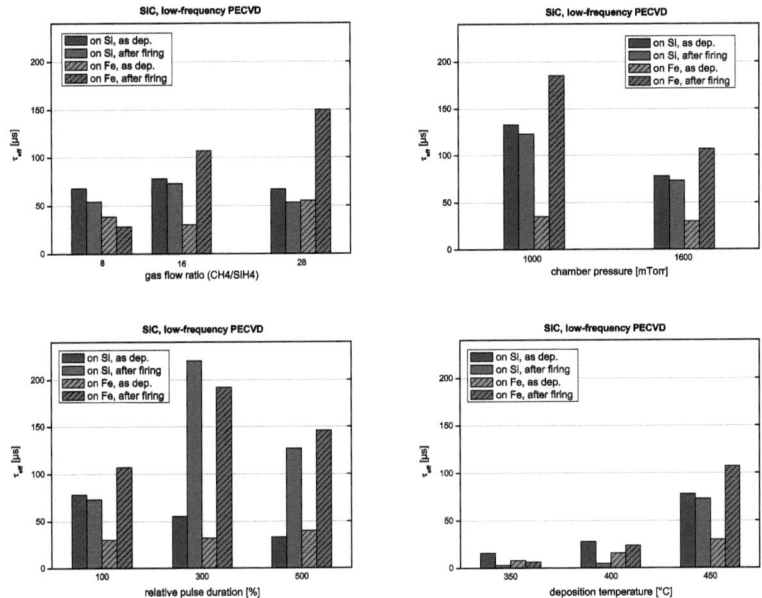

Figure 5.7: *Effective lifetime results of the first series of SiC_x experiments with low frequency on p-type FZ wafers, as-deposited and after firing, located either on a Si wafer (Si) or a stainless steel support (Fe, only supporting the wafer on its edge) as substrate during deposition. Values are from single wafers. Standard values: $CH_4:SiH_4=16$, pressure=1600 mTorr, relative pulse duration=100%, deposition temperature=450° C*

5.3.2 SiC_x: surface passivation dependence on deposition parameters

To characterize the passivation performance, effective lifetimes were measured as-deposited and after firing (same equipment and parameters as used for metal-paste co-firing). All investigated parameters turned out to influence the passivation quality of the layers. In the following, they are discussed in increasing order of significance, relative to the range they were investigated in.

Gas flow ratio

While no clear influence of the $CH_4:SiH_4$ gas flow ratio is visible for samples placed on the Si support and all lifetimes decrease slightly due to the firing, there is an increase in lifetime after firing for samples placed on the Fe support, with lifetimes increasing with gas flow ratio. The overall influence of the gas flow ratio is very low as compared to SiC deposited with high-frequency PECVD (chapter 5.2) and [Martin03, Ferre08], and the trend of increasing lifetimes with increasing carbon content is opposite.

Chamber Pressure

A chamber pressure decrease from 1600 to 1000 mTorr causes a clear increase in surface passivation for both support types, while the trend of slightly decreasing passivation after firing for the Si support and strongly increasing passivation for the Fe support is the same as for the gas flow ratio.

Power pulsing ratio

Surface passivation after firing reached an optimum at 300% relative pulse time, while lifetimes before firing decreased on the Si support and remained almost unchanged on the Fe support.

Deposition temperature

Substrate temperature has a significant influence on the surface passivation both before and after firing, for either support type. Again, samples from the Fe support benefit from firing except for the lowest temperature of 350°C, while samples on the Si support deteriorate in surface passivation. There is a clear increase in surface passivation with temperature, with the relative optimum at the highest temperature investigated, pointing to a potential absolute optimum at even higher temperatures. This was investigated and confirmed in a subsequent study (section 5.4). The observed positive influence of deposition temperature is the same as for high-frequency PECVD [Martin03, Ferre08].

It could not be clarified whether the different behavior on Si and Fe support and generally lower surface passivation from the Si support was

- due to inflicted surface impurities or damage of the wafer side resting on the Si support during deposition on the first side, or

- due to the differences in the electric field strength over the wafer in the plasma, as wafers on the Fe substrate always had higher film thickness.

5.3.3 SiC_xN_y: surface passivation dependence on deposition parameters

To characterize the passivation performance, effective lifetimes were measured as-deposited and after firing (same equipment and parameters as used for metal-paste co-firing). All investigated parameters turned out to influence the passiva-

5.3. Low Frequency Direct Plasma

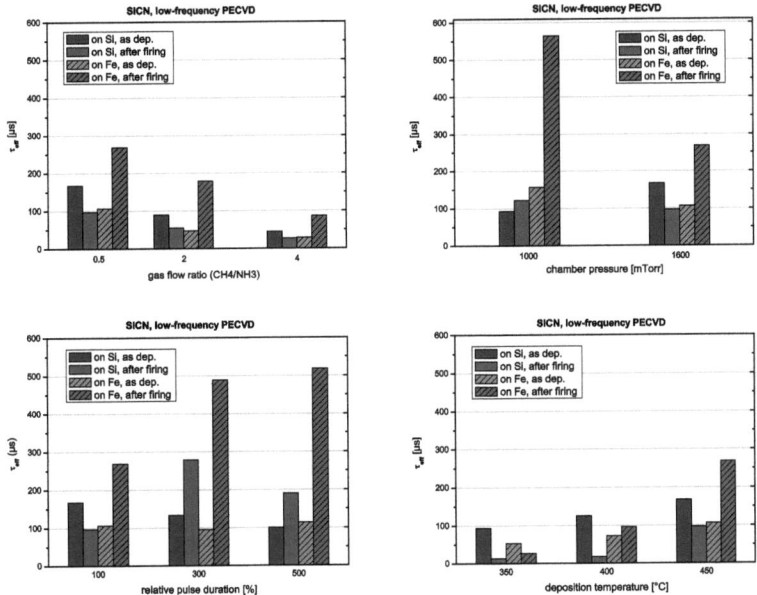

Figure 5.8: *Effective lifetime results of the first series of SiC_xN_y experiments with low frequency on p-type FZ wafers, as-deposited and after firing, located either on a Si wafer (Si) or a stainless steel support (Fe, only supporting the wafer on its edge) as substrate during deposition. Values are from single wafers. Standard values: $CH_4:NH_3=0.5$, pressure=1600 mTorr, relative pulse duration=100%, deposition temperature=450° C*

tion quality of the layers. In the following, they are discussed in increasing order of significance, relative to the range they were investigated in.

Gas flow ratio

The $CH_4:NH_3$ gas flow ratio has a clear impact on surface passivation both before and after firing for either support type. The influence of increasing methane concentration is clearly detrimental, as opposed to the results for SiC_x when regarding methane to silane ratio. This result is in contradiction to [Ferre08], but this author used only P-doped SiC_xN_y films, and also found that the position parameter dependence was different for P-doped and intrinsic SiC_x. The beneficial influence of lower C-content on surface passivation found in this work was also observed in [Martin03], where intrinsic SiC_xN_y was deposited using N_2 instead of NH_3. There, nitrogen addition yielded a sixfold increase in lifetime as compared to SiC_x deposited with the same parameters (without N_2 flow), independent of the relative nitrogen gas flow.

Even though no ellipsometer was available, the colour intensity of the deposited films pointed to a clearly lower refractive index for all SiC_xN_y films in this study compared to the SiC_x films, hinting rather to C-doped SiN_x than to SiC_xN_y, especially for the lowest methane/ammonia ratio of 0.5, where surface passivation comparable to SiN_x (see chapter 4) after firing was reached with decreased chamber pressure.

This is consistent with the observation that while a refractive index of 1.9 (N:Si ratio in the layer of slightly below 1 [Romijn06]) can be reached with the Centrotherm system with a NH_3:SiH_4 gas flow ratio of 10:1 for SiN_x (chapter 4), 35 times as much methane as silane is needed to achieve a refractive index of 2.4 after firing (C:Si ratio just slightly below 1 according to [Bullot87]) for SiC_x (section 5.4), pointing to much lower splitting rates of CH_4 compared to NH_3 in this system at comparable settings of plasma power and temperature.

Chamber Pressure

A chamber pressure decrease from 1600 to 1000 mTorr causes a clear increase in surface passivation for the Fe support both before and after firing, while for the Si support the initially higher lifetime at higher pressure degrades more during firing, such that the lifetime after firing is also higher at lower pressure for the Si support. The trend of strongly increasing passivation by firing for the Fe support is the same as for the gas flow ratio.

Power pulsing ratio

Surface passivation after firing reached a relative optimum at 500% relative pulse time on the Fe support, while lifetimes on the Si support decreased before firing and showed a maximum for 300% relative pulse time after firing. Lifetimes from samples on Si support clearly improved due to firing with increased pulse time, as opposed to 100%, where firing reduces lifetime.

Deposition temperature

As with SiC_x, the substrate temperature has a significant influence on the surface passivation both before and after firing, for either support type. Again, samples from the Fe support benefit from firing except for the lowest temperature of 350°C, while samples on the Si support deteriorate in surface passivation due to firing. There is a clear increase in surface passivation with temperature, with the relative optimum at the highest temperature investigated, pointing to a potential absolute optimum at even higher temperatures. The observed positive influence of deposition temperature is the same as for SiC_x and SiC_xN_y from high-frequency PECVD [Martin03, Ferre08].

5.3. Low Frequency Direct Plasma

It could not be clarified whether the different behavior on Si and Fe support and generally lower surface passivation from the Si support was

- due to inflicted surface impurities or damage of the wafer side resting on the Si support during deposition on the first side, or
- due to the differences in the electric field strength over the wafer in the plasma, as wafers on the Fe substrate always had higher film thickness.

5.3.4 p^+-passivation

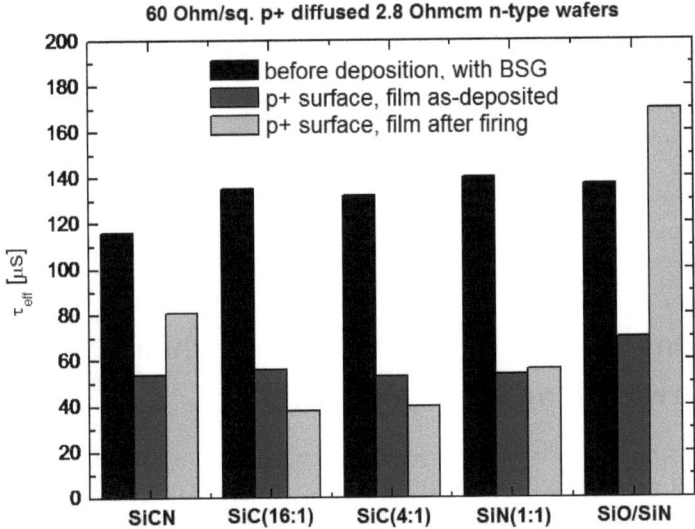

Figure 5.9: p^+ passivation by different layers of SiN_x, SiC_x and SiC_xN_y. Numbers in brackets denote gas flow ratios of (methane:silane) for SiC_x and (ammonia:silane) for SiN_x

For p^+ passivation comparison, the following layers were deposited on 2.8 Ωcm n-type Cz material with double-sided 60 $\Omega/\square$ BBr$_3$-diffusion [Petres05a]:

- The best-passivating layer of SiC_x on the Si support from the comparison of gas flow ratios (CH$_4$:SiH$_4$ GFR=16) (samples could not be deposited on a Fe support this time)

- A SiC_x layer with an especially low gas flow ratio (4 instead of 8 which was the lowest value in the study)

- The best-passivating SiC_xN_y layer (with lowest methane/ammonia gas flow)

- Si-rich SiN_x (GFR=1)

- A stack of 10 nm thermal SiO_2 followed by a top layer of 60 nm SiN_x (GFR=1)

The reason for including the SiN_x and SiC_x layer with especially low GFR in this study is that layers of both SiN_x and SiC_x with highest Si-content have consistently been found to contain the lowest (positive) fixed surface charge density Q_f [Martin03, Kerr03, Ferre08]. A very low positive Q_f was deemed to be a key factor for improving p^+-surface passivation [Kerr03, Libal05].

The boron silicate glass (BSG)-passivated surface was used as reference value. Fig. 5.9 shows the resulting effective lifetimes at 1 sun illumination of the investigated films before and after firing, as compared to the BSG-passivated surface.

All investigated passivation schemes except for the fired thermal-SiO_x/SiN_x stack performed lower than the BSG. The injection-level dependences of the lifetime curves are similar to the results in [Vetter06] with a local maximum below Auger-limited levels and a local minimum at lower injection levels than the local maximum. This suggests a positive Q_f still too high for sufficient surface passivation.

5.4 DoE on gas flow ratio, power, chamber pressure and temperature

To investigate the influence of the temperature, chamber pressure, plasma power density and gas flow ratio on the passivation and optical properties of the deposited SiC_x films in a broad range, a multilevel-factorial DoE was applied, using DoE software to obtain the maximum information from a certain number of different parameter combinations and thus deposition runs. The depositions were carried out with the Centrotherm system at ISC Konstanz, using the horizontal boat.

As the deposition rate normally increases with the plasma power P, the deposition time was scaled linearly with $1/P$, thus prolonged for the 1600 W and shortened for the 4000 W depositions compared to the standard 600 s for 2800 W, to compensate the thickness difference at least in part. Later, the thickness was recalculated for a deposition time of 600 s to allow direct growth rate comparison in figure 5.11. This recalculation is straightforward as except for the very first monolayers, film deposition rate in the PECVD is not time-dependent for a given parameter set [Per03], which was also confirmed within this work.

Again, the passivation quality was assessed by measuring effective lifetimes at 1 sun illumination by QSSPC on p-, n- and p^+-Si surfaces, both as-deposited and

5.4. DoE on gas flow ratio, power, chamber pressure and temperature

after the firing step. The chamber pressure, substrate temperature, plasma power and gas flow ratio were varied, as they have been found to be significant both in the first LF-PECVD experiments and for HF-PECVD SiC_x-layers [Martin03, Ferre08, Riegel08].

The results of the multilevel-factorial DoE were evaluated to find the parameters for highest effective lifetimes both as deposited and after firing. The thickness of the studied SiC_x layers was above 50 nm, to ensure that the surface passivation is not limited by the film thickness (as mentioned before, a film thickness above 30-40 nm was found to be sufficient to avoid thickness-related limitation of the passivation quality, regardless of the layer composition [Martin03, Kerr03, Ferre08]).

For comparison, PECVD-films of SiN_x normally used for emitter passivation and single layer anti-reflective coating (ARC) were deposited as well as a recently developed SiO_xN_y, both as single layer ARC and in stack with the aforementioned SiN_x (ca. 10nm SiO_xN_y + 70 nm SiN_x).

Table 5.3 lists the investigated parameters and their range and number of levels of variation.

Factors	Range	Levels
Gas flow ratio Y_C(CH_4/SiH_4)	10 - 35	4
RF power P[W]	1600 - 4000	3
Pressure p[mTorr]	1400 - 2000	3
Temperature T[°C]	450 - 500	2

Table 5.3: *DoE matrix used to investigate the passivation quality of SiC_x layers deposited in the Centrotherm low-frequency PECVD system at ISC Konstanz*

The depositions were carried out on 1.5 Ωcm n- and 4 Ωcm p-type Cz-wafers, both on the base-doped samples and on top of both-sided boron diffusion of 65 $\Omega/\square$. Each set of two subsequent depositions (front and back side, to obtain symmetrical lifetime samples) featured two samples of the same structure, resulting in 8 wafers receiving the same depositions.

All samples were measured as deposited with μ-PCD and subsequently QSSPC within the area of the highest effective lifetime according to the μ-PCD lifetime map. The QSSPC-determined effective lifetimes were taken both at 1 sun illumination and at an injection level of 10^{15} cm^{-3}. Four out of the eight wafers per run, i.e. one out of each set of two identical samples, were fired after the lifetime measurements and subsequently re-measured. The other four were kept in their as-deposited state to have references for possible later comparisons of the effective lifetimes as well as the optical and chemical properties of the layers. To determine the latter two, ellipsometry and FTIR measurements were carried out.

5.4.1 General results

The optical and thickness homogeneity of the deposited layers was good, with thickness variations in between two samples of the same group below 3 % according to the ellipsometry measurements, and no variations in the refractive index were detectable.

The main difference to the SiC_x depositions from high-frequency PECVD, was the much lower influence of the precursor gas flow ratio of methane to silane. It turned out that the bulk of the BBr_3-diffused n-type wafers, although having received identical pre-treatment as the p-type wafers, had degraded strongly after diffusion (τ_{eff} with BSG below 10 µs), thus rendering them worthless for the study. The presented results on p$^+$ are therefore solely from the BBr_3-diffused p-type wafers.

5.4.2 Gas flow ratio and plasma power dependence

According to the pareto chart for undiffused p-type, the gas flow ratio has the strongest influence on the effective lifetime before firing (BF, i.e. as-deposited), followed by the plasma power. The influences of the chamber pressure and temperature alone are below the 5% significance level, and therefore not considered to be relevant.

An overall optimum for the lifetime before firing was predicted by the DoE outside the parameter hypercube for even lower power, lower pressure and higher temperature. Thus, an additional parameter set was investigated with P=1100 W, p=1200 mTorr and T=530°C. It yielded indeed a higher effective lifetime before firing of 315 µs at one sun (350 µs at 10^{15} cm^{-3}) which is slightly lower than standard SiN_x on the same material with 330 µs (480 µs at 10^{15} cm^{-3}). This SiC_x layer degraded to 140 µs after firing.

The predicted optimum after firing was not investigated as only a moderate improvement compared to the already realized parameter sets was expected. The observed tendency of the lifetimes before and after firing fits the results obtained with high-frequency PECVD: while lifetimes continue to increase with refractive index (and thus presumably with Si-content) before firing, there is an optimum after firing at n=2.6-2.7 (see figure 5.11 bottom). A possible explanation for this optimum after firing is offered in chapter 5.6.

A possible explanation for the decrease of the refractive index after firing for films with n<2.9 is the out-diffusion of hydrogen from the films leaving nanometer-sized voids, counter-acted by a film-condensing mechanism that increases with refractive index and thus silicon content of the films. However, no explanation for this hypothetical mechanism can be offered as there is no observable decrease in film thickness to account for physical densification. Interestingly, the observed trends are contrary to those found for SiC_x deposited by high-frequency remote plasma PECVD [Janz06], but would be almost identical

5.4. DoE on gas flow ratio, power, chamber pressure and temperature

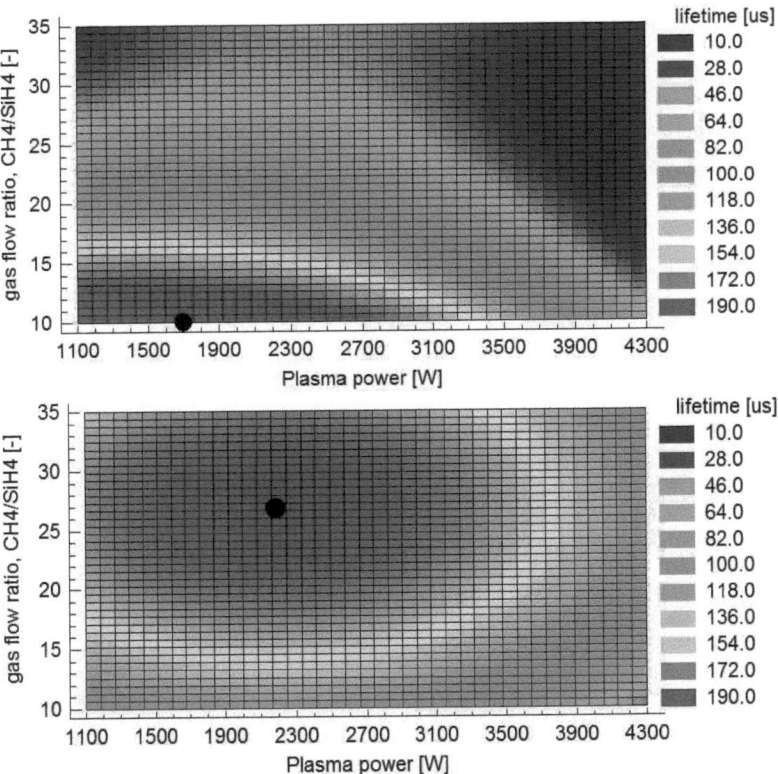

Figure 5.10: *Influence of gas flow ratio and plasma power on the passivation properties of SiC_x as deposited (top) and after firing (bottom) by low-frequency PECVD on p-type Cz wafers at pressure p=1400 mTorr and temperature T=500° C. Values are averages of two wafers for lifetimes before firing and from single wafers for lifetimes after firing, actual values from depositions with eight different parameter combinations and interpolated values for the remaining six parameter combinations. The black dots mark the optimum sets of parameters as predicted by the DoE software for this pressure and temperature. While lifetimes on n-type material were slightly lower for some parameter sets and higher for others compared to p-type, the overall distribution and trends are comparable and therefore not shown separately.*

if the values for as-deposited and annealed samples were exchanged. As it can be excluded that samples were mixed up in this work, no explanation can be given for this phenomenon.

Chapter 5: PECVD-Silicon Carbide and Silicon Carbonitride

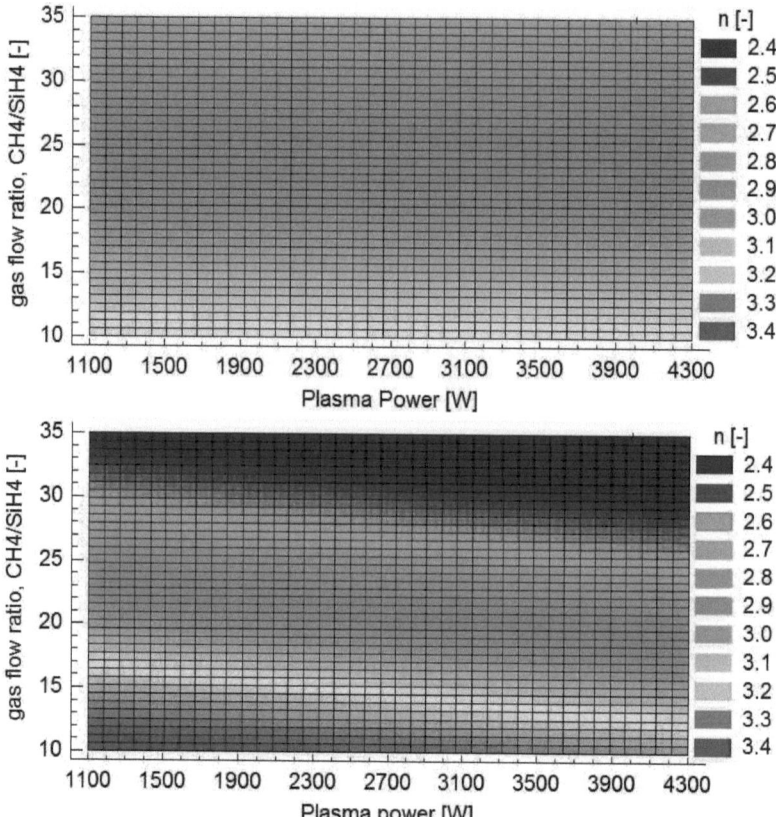

Figure 5.11: *Influence of gas flow ratio and plasma power on the refractive index of SiC_x as deposited (top) and after firing (bottom) by low-frequency PECVD on p-type Cz wafers at pressure $p=1400$ mTorr and temperature $T=500°C$. Values are from single wafers, actual values from depositions with eight different parameter combinations and interpolated values for the remaining six parameter combinations.*

The suggested partial formation of μc-Si and μc-C [Palma99] that offers an explanation for the observed passivation trends (see chapter 5.4) would imply refractive index changes that are opposite to what is observed: while the refractive index of C at 633 nm increases with crystallisation (a-C has about 1.9, graphite 2.2 and diamond 2.4), it decreases for Si (a-Si has 4.52, c-Si 3.89)[Sopra10]. Thus, the refractive index should be increased by firing for C-rich SiC_x and decreased for Si-rich SiC_x.

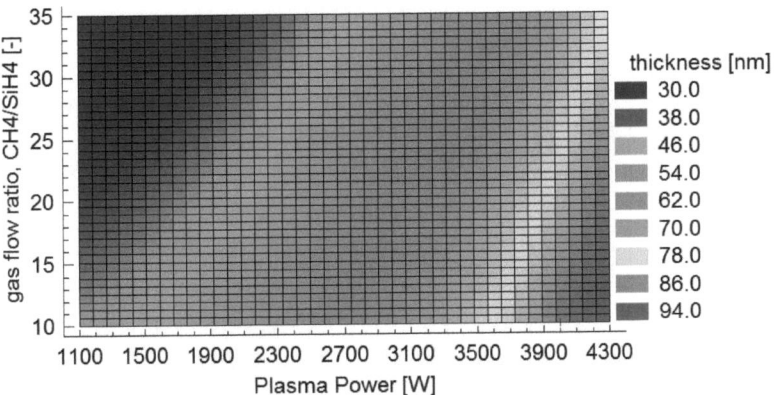

Figure 5.12: *Influence of gas flow ratio and plasma power on the film thickness of SiC_x deposited by low-frequency PECVD after firing. Deposition time was kept constant at 600 s, pressure and temperature at $p=1400$ mTorr and $T=500°C$. Values are from single wafers, actual values from depositions with eight different parameter combinations and interpolated values for the remaining six parameter combinations. Thicknesses before firing were almost identical (below 3% variation which corresponds to the variation between samples) and are therefore not shown.*

SiC_x deposited thicknesses depending on gas flow ratio and plasma power measured after firing show the same trends as SiN_x deposited from SiH_4 and NH_3 in the same machine. The dependency on the gas flow ratio is weaker than for SiN_x, while the dependency on power is stronger. The most likely explanation for this behavior which is consistent with the observed dependencies of the refractive index on gas flow ratio and power is the higher dependency of CH_4 splitting on the availability of SiH_4 radicals of sufficient energy as compared to NH_3 splitting, due to the higher bonding and thus splitting energy of C-H bonds in comparison to N-H bonds. The same trend and explanation are found in [Laveuve07].

5.4.3 p^+-passivation

As in the previous study (section 5.3), the best SiC_x films found in the DoE study in terms of surface passivation before and after firing and the SiO_xN_y/SiN_x stack with 10 nm SiO_xN_y were also tested on p^+. This time, the 4 Ωcm p-type samples received 90 $\Omega/\square$ BBr_3-diffused surfaces (measured on the n-type material). The results are shown in fig. 5.13.

Similar to the results with HF-PECVD in section 5.3.4, the achieved p^+ surface passivation quality is rather low, and only the SiO_xN_y/SiN_x stack performs comparable to the BSG (≈ 100 μs, not shown). Again, the reason is likely due to

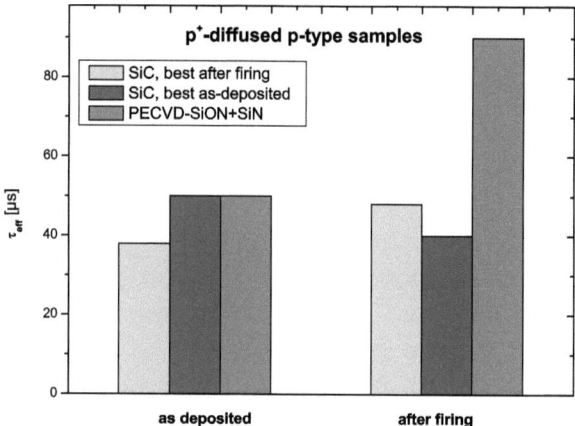

Figure 5.13: p^+ passivation by the best SiC_x films for surface passivation before (red) and after firing (yellow) compared to the SiO_xN_y/SiN_x stack

the fixed positive surface charge, whose influence is visible [Martin03, Kerr03] in the injection level dependent comparison of SiC_x layers before and after firing on p- and n-type in fig. 5.14: While the τ_{eff} continually increases with decreasing Δn for n-type and the higher lifetimes before and after firing are from the same samples, a maximum is reached on p-type with the C-rich SiC_x, and this sample performs lower than the one covered with Si-rich SiC_x after firing which shows a plateau on the lower end of the measurable injection level range. This points to a fixed positive surface charge density on both films, with the higher charge density found in the C-rich SiC_x. This is consistent with the findings of [Martin03, Kerr03, Ferre08].

5.4.4 Preplasma dependence

At the beginning of the DoE, a CH_4-preplasma was used prior to depositing SiC_x from CH_4 and SiH_4 for the first depositions in analogy to the SiN_x depositions where a NH_3 preplasma is used for surface-cleaning prior to adding SiH_4 to the reaction chamber and thus initializing the SiN_x deposition reaction.

When said first SiC_x depositions yielded poor passivation quality both before and after firing, it was assumed that the use of CH_4 instead of NH_3 may be the cause, as CH_4 should have caused deposition of a hydrogenated amorphous carbon (a-C:H) film. An investigation was carried out by repeating depositions with the same parameter sets twice more, but once with an NH_3 preplasma and once with no preplasma at all. The following figure 5.15 shows the results

5.4. DoE on gas flow ratio, power, chamber pressure and temperature

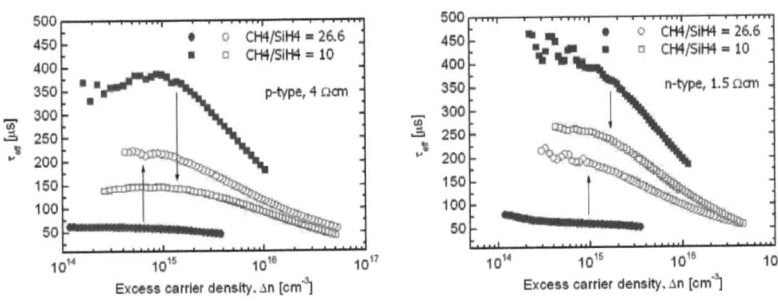

Figure 5.14: *Effective lifetime as a function of excess carrier density measured on p-type and n-type wafers for two different gas flow ratios. The measurements are performed before firing (filled symbols) and after firing (empty symbols)*

of these variations with the parameter set that yielded best surface passivation before firing.

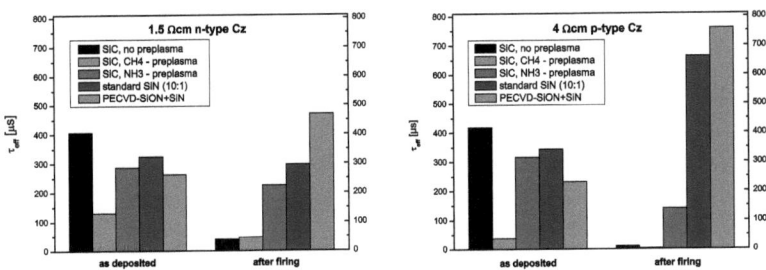

Figure 5.15: *Influence of preplasma on the passivation properties of SiC_x deposited by low frequency PECVD before and after firing, compared to standard SiN_x and a SiO_xN_y/SiN_x stack with 10 nm SiO_xN_y. Values are effective lifetimes at one sun illumination and from single wafers.*

It is clearly visible that the preplasma has a strong effect on the passivation properties before and after firing. While the film deposited without preplasma exhibits best passivation before firing, both on n- and p-type Si compared to all other layers, it degrades strongly through the firing step. The film deposited with CH_4 preplasma shows low passivation quality before firing that almost vanishes completely after firing on p-type Si and degrades further on n-type Si, in either case to the same level as the SiC_x deposited without preplasma, most likely because of a thin layer of a-C formed at the interface. Best SiC_x passivation after firing comes from the SiC$_x$ deposited after NH$_3$ preplasma cleaning, while the overall best values on n- and p-type Si result from the SiO$_x$N$_y$/SiN$_x$ stack with 10 nm SiO$_x$N$_y$.

5.5 FTIR-study of SiC$_x$ layers from low-frequency PECVD

bond	peak/band range [cm^{-1}]	bond	peak/band range [cm^{-1}]
N-H	3340	C-H	1110
C-H$_3$	2960	Si-O	1105
C-H$_2$	2880	C-H$_n$	1100-950
O=Si-H	2400-2300	N-H$_n$	1100-900
Si-H$_3$	2140	Si-CH$_2$	990
Si-H$_2$	2090	Si-N	850, 810
Si-H	2000	Si-CH$_3$	780
(Si-H$_2$)$_n$	2090-2000	C-H	760
H$_2$O	1900-1300	Si-C	670
Si-CH$_3$	1250	Si-H	615

Table 5.4: *List of absorption peaks and related bonds used for IR spectra evaluation of the SiN$_x$, SiC$_x$ and SiC$_x$N$_y$ films. The precise peak position of the Si-H peaks between 2000-2200 cm^{-1} is typically varying based on the film stoichiometry, e.g. a more C-rich film will cause peaks to be shifted towards higher wavenumbers due to the higher bonding energy of C compared to Si. The same effect is observable for higher N-contents.*

The films from the DoE in section 5.4, varying plasma power, gas flow ratio, chamber pressure and temperature, were also investigated by fourier-transform infrared spectroscopy (FTIR). Table 5.4 lists the observed absorption peaks and their corresponding wavenumbers. The absorption intensities were normalized to a film thickness of 100 nm to allow direct comparison, as the different investigated deposition parameter sets yielded film thicknesses between 32-112 nm. As the intensity of the peaks around 1200-600 cm^{-1} was typically more than one order of magnitude higher than those between 3500-1200 cm^{-1}, the IR spectra are displayed in two separate figures with accordingly different absorption intensity scales.

Figure 5.16 and 5.17 show the IR-spectra of SiN$_x$ deposited at standard parameters and with three different gas flow ratios of NH$_3$:SiH$_4$ for comparison. The layers were characterized after firing. As with all FTIR curves in this work, the

5.5. FTIR-study of SiC$_x$ layers from low-frequency PECVD

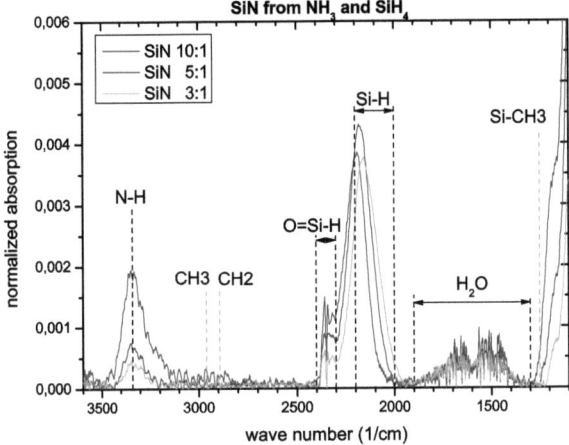

Figure 5.16: *FTIR spectra (small peak regime) of SiN$_x$ layers deposited by low-frequency PECVD with different (ammonia:silane) gas flow ratios. C-related peaks are shown for later comparison with SiC$_x$ and SiC$_x$N$_y$ in figures 5.18 to 5.21*

results are displayed in separate graphs for the "small peak regime" (3600..1100 cm^{-1}) and the "large peak regime" (1300..500 cm^{-1}), and peak intensities are normalized to a film thickness of 100 nm for best comparability.

The Si-N bond density decreases with decreasing NH$_3$/SiH$_4$ gas flow ratio and thus decreasing N content of the films as could be expected.

The absolute values of the Si-N bond densities for the three investigated SiN$_x$ films (see caption of figures 5.16 and 5.17) are in the range of the ones in [Romijn05] and fit the previously reported tendency that higher Si-N bond densities up to $1.3 \cdot 10^{22}$ cm^{-3} correspond to increasing surface passivation quality [Mäckel02, Romijn05].

While also N-H bond density decreases with decreasing NH$_3$/SiH$_4$ gas flow ratio and thus decreasing N content of the films, the Si-H bond density between 2200-2100 cm^{-1} remains mostly constant. A change is however observed for the position of the peak which is shifted towards lower wavenumbers and thus lower bond energies by about 50 cm^{-1} with increasing Si content. This reveals a shift to more Si- than N-backbonding of the Si-H bonds [Romijn05] and is consistent with the concentration changes of Si and N. The same effect is reported for SiC$_x$ by [Ehling10] and is visible for the C-containing films in this work.

The difference in intensity of the H$_2$O-band "noise" from 1900-1300 cm^{-1},

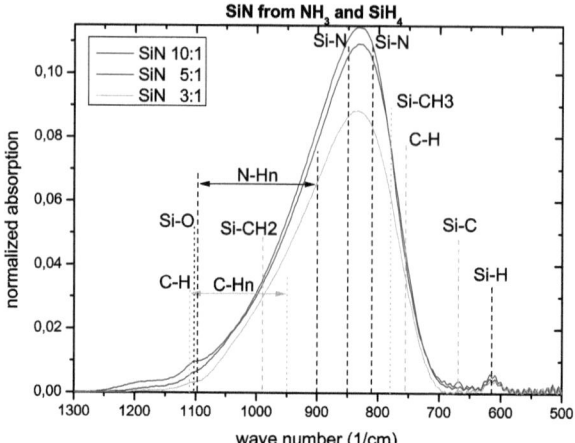

Figure 5.17: *FTIR spectra (large peak regime) of SiN_x layers deposited by low-frequency PECVD with different (ammonia:silane) gas flow ratios. Si-N bond densities are $10.4 \cdot 10^{22}$ cm^{-3} for the (10:1), $10.0 \cdot 10^{22}$ cm^{-3} for the (5:1) and $8.2 \cdot 10^{22}$ cm^{-3} for the (3:1) SiN_x. C-related peaks are shown for later comparison with SiC_x and SiC_xN_y in figures 5.18 to 5.21*

originating from adsorbed water at the sample surface, is due to the thickness normalization of the curves causing higher intensities for thinner films. The same is true for the Si-O peak from interfacial SiO_x at 1105 cm^{-1}.

For the SiC_x layers investigated within the DoE in section 5.4, the comparison of as-deposited (BF) and after-firing (AF) layer composition in figures 5.18 to 5.21 does not provide information to allow a quantitative prediction of the level of surface passivation that a layer is providing before firing and how much it is affected by firing. However, a relation to predict the surface passivation quality and trend is found: a visible decrease in Si-H peak intensity after firing typically corresponds to improved surface passivation after firing, indicating H-atoms being released for surface passivation. This is consistent with results of [Romijn05, Cic07, Shir09] for SiN_x films. The total peak height is related to the overall level of surface passivation. For firing-stable C-rich SiC_x layers, an additional decrease in C-H2, at or slightly below 2880 cm^{-1}, is observable, as well as for the other C-H2 appearance in the C-Hn band between 1100 and 950 cm^{-1}. In contrast to this, the C-H peak at 770 cm^{-1} generally increases after firing, indicating that only one hydrogen atom is typically released from C-H2 bonds, while more C-H bonds are created than broken up, if any, as their bonding energy is even higher

5.5. FTIR-study of SiC$_x$ layers from low-frequency PECVD

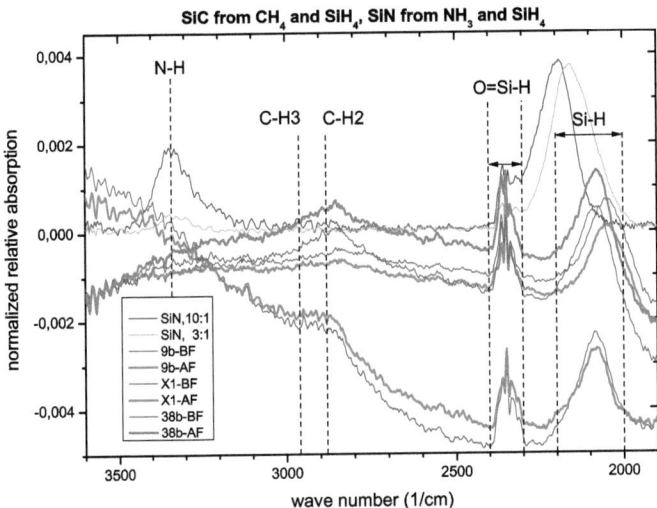

Figure 5.18: *FTIR spectra (small peak regime) of different SiC$_x$ layers deposited at 500° C, giving best surface passivation before firing (X1, n_{AF}=3.0), best surface passivation after firing ("38b", n_{AF}=2.8), and second lowest refractive index before and after firing ("9b", n_{AF}=2.5). SiN$_x$ with GFR 10 and 3 is shown for comparison.*

than that of N-H, which rarely contribute to H released from SiN$_x$ [Romijn05].

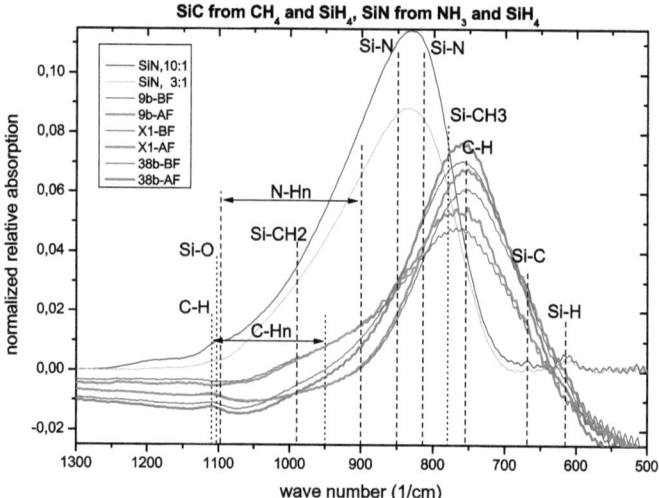

Figure 5.19: *FTIR spectra (large peak regime) of different SiC_x layers deposited at $500°C$, giving best surface passivation before firing (X1, $n_{AF}=3.0$), best surface passivation after firing ("38b", $n_{AF}=2.8$), and second lowest refractive index before and after firing ("9b", $n_{AF}=2.5$). SiN_x with GFR 10 and 3 is shown for comparison.*

5.6 Comparison of a-SiC$_x$:H to a-SiN$_x$:H

The different tendencies observed for SiN$_x$ and SiC$_x$, mainly regarding the dependency of surface passivation quality before and after firing and on the refractive index, can possibly be explained as follows:

- a-SiC$_x$ has been observed to typically exhibit phase-segregation into microcrystalline silicon (μc-Si) and microcrystalline graphite (μc-C) when heated, whereas μc-SiC would only form when the ratio of Si and C is nearly stoichiometric (i.e. ≈1) [Palma99]. This may explain the observed optimum in firing stability within the Doe with low-frequency PECVD for layers with n≈2.7-2.8, which is near the refractive index of (stoichiometric) c-SiC of 2.65-2.69 [Weimer97].

- a-SiC$_x$:H was found to be less dense with increasing C-content [Mahan86, Ehling10] which can result from higher methane content in the precursor gas mix, higher plasma power density or chamber pressure or a combination of these factors. The lower film density also allows for easier out-diffusion of hydrogen from the films during the firing step, resulting in a decrease in refractive index, which was observed in section 5.4. This is comparable to

5.6. Comparison of a-SiC$_x$:H to a-SiN$_x$:H

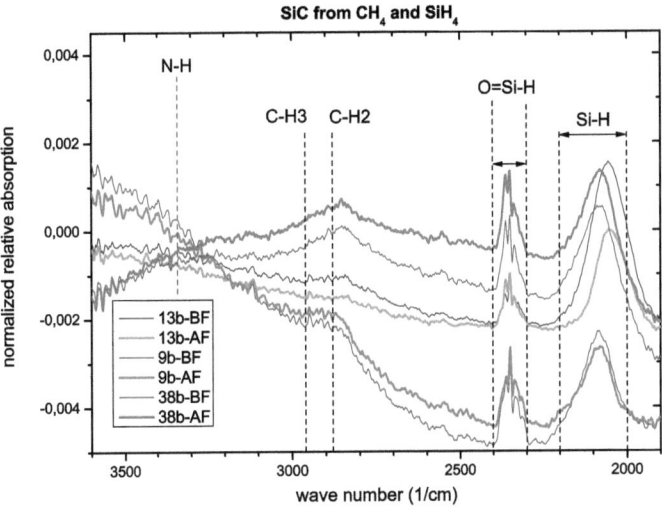

Figure 5.20: *FTIR spectra comparison of different SiC$_x$ layers deposited at 500° C, with highest pressure ("9b", n_{AF}=2.5/"13b", n_{AF}=3.1) and lowest plasma power ("13b", n_{AF}=3.1/"38b", n_{AF}=2.8).*

a-SiN$_x$:H, where layers of lower refractive index (below n≈2.2 in this work) exhibit a decrease in refractive index after firing, while layers of higher n increase in n after firing. This is also observed for a-SiC$_x$:H of refractive indices above 2.8 (section 5.4) which show an increase in refractive index after firing for the lower of the investigated plasma power densities.

- While a-Si:H can yield excellent surface passivation, a-C:H shows almost no surface passivation of c-Si, which is also visible from the preplasma variation experiment at the end of chapter 5.3. This is in good agreement with the general observation that increasing the relative C-content above 50% is always detrimental for surface passivation, independent of the precursor gases used for a-SiC$_x$:H or a-SiC$_x$N$_y$:H deposition. An analogous comparison with nitrogen would only be possible at extremely low temperatures due to the boiling point of nitrogen of -196°C. However, results of decreasing surface passivation of layers with n<1.9 in [Aberle99] and [Kerr03] also point to a detrimental influence of nitrogen roughly above 50% relative content in a-SiN$_x$:H. The optimum in terms of refractive index of the compound when regarding surface passivation quality after firing is higher for a-SiC$_x$:H than for a-SiN$_x$:H, as layers of identical stoichiometry have a higher refractive index for a-SiC$_x$:H (2.6-2.7) than for a-SiN$_x$:H (1.9-2.0). The higher refractive index for a given stoichiometry for SiC$_x$ as compared

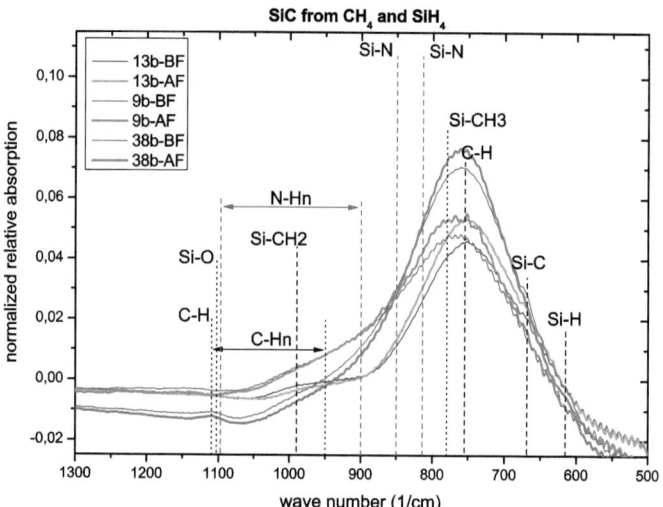

Figure 5.21: *FTIR spectra comparison of different SiC_x layers deposited at 500° C, with highest pressure ("9b", $n_{AF}=2.5$/"13b", $n_{AF}=3.1$) and lowest plasma power ("13b", $n_{AF}=3.1$/"38b", $n_{AF}=2.8$).*

to SiN_x corresponds to the higher refractive index of a-C of 1.9 [Sopra10] compared to that of (liquid) nitrogen of 1.2 [Bar10].

Bibliography

[Anger08] H. Angermann et al., *Wet-chemical passivation of atomically flat and structured silicon substrates for solar cell application*, Appl. Surf. Sci.**254** 12, pp. 3615-3625 (2008)

[Aru06] J. Arumughan et al., *Application of BTBAS based silicon nitride in buried-contact screen-printed PERT-type bifacial solar cells*, Proc. 21st EU-PVSEC, Dresden, Germany (2006)

[Bar10] K Barbalace, *Periodic Table of Elements*, EnvironmentalChemistry.com. 1995 - 2010. internet source: http://EnvironmentalChemistry.com/yogi/periodic/N.html

[Ber98] C. Berge et al., *Comparison of effective carrier lifetime in silicon determined by transient and quasi-steady-state photoconductance measurements*, Proc. 2nd WCPEC, Wien, p.1426 (1998)

[Bullot87] J. Bullot and M. P. Schmidt, *Physics of Amorphous Silicon-Carbon Alloys*, Phys. Stat. sol. (b) **143**, 345 (1987)

[Cic07] J. Cichoszewski, *Silicon nitride antireflection and passivation layers for solar cells*, Diploma thesis, Institute for Physical Electronics, University of Stuttgart, Germany (2007)

[Ehling10] C. Ehling et. al., *a-SiC:H passivation for crystalline silicon solar cells*, Phys. Stat. Sol. C7, No. 3-4, 1016-1020 (2010)

[Ferre08] R. Ferré, *Surface passivation of crystalline silicon by amorphous silicon carbide films for photovoltaic applications*, PhD thesis, Universitat Politecnica de Catalunya, Barcelona, Spain (2008)

[Gabor10] A. Gabor et al., proc. 5th WCPEC, Valencia (2010) - *currently in print*

[Geer04] L.G.Geerligs et al., *Base Doping and Recombination Activity of Impurities in Crystalline Silicon Solar Cells*, Prog. Photovolt: Res. Appl. **12**, pp.309-316 (2004)

[Grohe03] A.Grohe et al., *Characterization of laser-fired contacts processed on wafers with different resistivities*, Proc. 3rd WCPEC, Osaka, Japan (2003)

[Henley09] F. Henley et al., *Kerf-Free 20-150 μm c-Si Wafering for Thin PV Manufacturing*, Proc. 24th EU-PVSEC, p.886, Hamburg, Germany (2009)

[Janz06] S. Janz, *Amorphous Silicon Carbide for Photovoltaic Applications*, Dissertation, Fraunhofer ISE, Freiburg/University of Konstanz, (2006)

[Kerr02] M. J. Kerr et al., *Lifetime and Efficiency Limits of Crystalline Silicon Solar Cells*, Proc. 29th IEEE PVSC, New Orleans, USA, (2002)

[Kerr02b] M. J. Kerr, *Surface, Emitter and Bulk Recombination in Silicon and Development of Silicon Nitride Passivated Solar Cells*, PhD thesis, Australian National University (ANU), Canberra, Australia (2002)

[Kerr03] M. J. Kerr et al., *Limiting efficiency of crystalline silicon solar cells due to Coulomb-enhanced Auger recombination*, Prog. Photovolt: Res. Appl. **11**(2), pp.97-104 (2003)

[Kita91] A. Kitagawa et al., *Influence of H_2O on the SiO_2 Growth by Parallel-Resonant RF Plasma Oxidation*, Jap. J. Appl. Phys.**30**(7B), pp. L1317-L1318 (1991)

[Lau97] T. Lauinger, A. Aberle, and R. Hezel, in Proc. 14th European Photovoltaics Solar Energy Conference (1997), p. 853.

[Laveuve07] C. Laveuve, *Untersuchungen amorpher Siliziumkarbidfilme für die Anwendung in kristallinen Siliziumsolarzellen*, Diploma thesis, ISFH Hameln/University of Hannover (2007)

[Mäckel02] H. Mäckel and R. Lüdemann, *Detailed study of the composition of hydrogenated SiNx layers for high-quality silicon surface passivation*, J. Appl. Phys. **92**(5), pp. 2602-2609 (2002)

[Mahan86] A. H. Mahan, P. Raboisson, and R. Tsu, Appl. Phys. Lett. **50**, 335 (1986).

[Martin01] I. Martín et al.,*Surface passivation of p-type crystalline Si by plasma enhanced chemical vapour deposited amorphous SiC_x:H films*, Appl. Phys. Lett. **79** (14), pp. 2199-2201 (2001)

[Martin02] I. Martín et al., *Surface passivation of n-type crystalline Si by plasma-enhanced-chemical-vapor-deposited amorphous SiC_x:H and amorphous SiC_xN_y:H* *films*, Appl. Phys. Lett. **81** (23), pp. 4461-4463 (2002)

[Martin03] I. Martín, *Silicon surface passivation by Plasma Enhanced Chemical Vapor Deposited amorphous silicon carbide films*, PhD thesis, Universitat Politecnica de Catalunya, Barcelona, Spain (2003)

[Mih08] V.D. Mihailetchi et al., *Nitric Acid Pre-treatment for the Passivation of Boron Emitters for N-type base Silicon Solar Cells*, Appl. Phys. Lett. **92**, 063510 (2008)

[Nist10] *National Institute of Standards and Technology Chemistry Webbook*, source: http://webbook.nist.gov/chemistry

[Palma99] C. Palma et al., *Laser-induced crystallization of amorphous silicon-carbon alloys studied by Raman microspectroscopy*, Appl. Surf. Sci. **138-139**, pp. 24-28 (1999)

[Per03] T. Pernau, *Impulse für die industrielle Produktion kristalliner Siliziumsolarzellen*, Dissertation, University of Konstanz, Germany (2003)

[Petres05a] R. Petres, *Untersuchungen der Passivierung von p^+-Siliziumoberflächen für die Herstellung von p^+n - Siliziumsolarzellen*, Diploma thesis, University of Konstanz (2005)

[Petres05b] R. Petres et al., *Passivation of p^+-Surfaces by PECVD Silicon Carbide Films - A Promising Method for Industrial Silicon Solar Cell Applications*, Proc. PVSEC-15, p.128, Shanghai, China (2005)

[Petres06] R. Petres et al., *Improvements in the Passivation of p^+-Si Surfaces by PECVD Silicon Carbide Films*, Proc. WCPEC-4, Waikoloa, Hawaii, USA (2006)

[Riegel08] S. Riegel, B. Raabe, R. Petres, S. Dixit, L. Zhou, G. Hahn, *Towards Higher Efficiencies for Crystalline Silicon Solar Cells Using SiC Layers*, Proc. 23rd EU PVSEC, Valencia, 1604-1607, 2008

[Romijn05] I. Romijn et al., *Passivating mc-Si solar cells using SiN_x:H: From magic to physics*, Proc. 20th EU-PVSEC, pp. 1352-1355, Barcelona, Spain (2005)

[Sachs08] E. Sachs, Massachusetts Institute of Technology (MIT), Boston, USA (2008) source: http://www.1366tech.com/v2/

[Schmi99] J. Schmidt, *Measurement of differential and actual recombination parameters on crystalline silicon wafers*, IEEE Trans. El. Dev. **46**(10), 2018 (1999)

[Schoe95]
M. Schöfthaler and R. Brendel, *Sensitivity and transient response of microwave reflection measurements*, J. Appl. Phys. **77**(7), 3162 (1995)

[Sin96]
R.A. Sinton and A. Cuevas, *Contactless determination of current-voltage characteristics and minority carrier lifetimes in semiconductors from quasi-steady-state photoconductance data*, Appl. Phys. Lett. **69**(17), 2510 (1996)

[Ske67]
E.R. Skelt and G.M. Howells, *The properties of plasma-grown SiO_2 films*

[Solbuz10] *2010 Marketbuzz Annual World Solar PV Industry Report*, Solarbuzz, San Francisco, USA (2010) source: http://www.solarbuzz.com/Marketbuzz2010-intro.htm

[Solsrv09] Source: http://www.solarserver.de/solarmagazin/solar-report_1108_e.html

[Sopra10] Source: http://www.sopra-sa.com/index2.php?goto=dl&rub=4

[Stem04] N. Stem and N.Cid, *Physical limitations for homogeneous and highly doped n-type emitter monocrystalline silicon solar cells*, Solid-State Electronics **48**, 2, (2004) [Tsao07] C. W. Tsao et al., *Low temperature bonding of PMMA and COC microfluidic substrates using UV/ozone surface treatment*, Lab Chip, 7, 499-505 (2007)

[Vetter06] M. Vetter, R. Ferré, I. Martín, P. Ortega, R. Alcubilla, R. Petres, J. Libal, R.Kopecek, *Investigation of the Surface Passivation of p^+-Type Si Emitters by PECVD Silicon Carbide Films*, Proc. 4th WCPEC, Waikoloa, 1271-1274, 2006 [Vig85] J. Vig et al., *UV/ozone cleaning of surfaces*, J. Vac. Sci. Technol. A, (1985)

[Weimer97] A. Weimer, *Carbide, nitride, and boride materials synthesis and processing*, Chapman&Hall, ISBN 0412540606 (1997)

[Zhu05] H. Zhu et al., *Thermal nitridation kinetics of silicon wafers in nitrogen atmosphere during annealing*, Thin Solid Films **474**, Issues 1-2, p. 326-329 (2005)

Summary

Chapter 1 gives an introductory overview of the current status of photovoltaics, with focus on crystalline silicon (c-Si) based technology. An essential contribution to the reduction of electricity generation costs at the solar module production level is to be expected mainly from reduced silicon consumption by using thinner wafers and/or employing cheaper silicon feedstock ("solar grade silicon"). Together with sufficient light trapping, the key factor to being able to exploit the combined cost reduction potential by using thin solar grade silicon wafers is the availability of very good and industrially applicable electronic surface passivation methods. That way, material quality makes almost no difference anymore when going down to wafer thicknesses of 30 μm from the current 150-200 μm. This gives the motivation for this work.

Chapter 2 gives an overview of the theoretical bases of surface passivation and antireflection coating and describes the methods and equipment used to characterize the layers created in this work. While surface passivation is quantified by the effective surface recombination velocity S_{eff}, this parameter cannot be measured directly. Instead, the lifetime measurements by QSSPC and μPCD carried out for this work give the effective minority carrier lifetime τ_{eff}. With certain simplifying assumptions, an upper limit for S_{eff} can be calculated solely from τ_{eff} and the sample thickness. As shown in chapter 2.2.1, the error resulting from this simplified approach often found in literature is not negligible for good surface passivation layers, but acceptable in practice as the focus is on comparing different passivation layers.

While the μPCD was applied to obtain spatially resolved lifetime maps of the entire sample, the QSSPC was subsequently used to determine absolute values of the best areas that can be compared with the literature, as QSSPC is the established standard in c-Si photovoltaics.

The refractive index and thickness of the investigated dielectric films were measured by spectroscopic ellipsometry, and the chemical composition was analyzed by Fourier-Transformed Infrared Spectroscopy (FTIR) to investigate relations with the surface passivation and optical properties.

Chapter 3 explains the difference between growth and deposition as methods to obtain a film on top of a substrate and gives some examples of the dielectrics and their formation techniques which are most commonly used in current c-Si solar cell technology, with focus on SiO_x and SiN_x from thermal oxidation and Plasma Enhanced Chemical Vapor Deposition (PECVD), respectively.

Subsequently, the thin film formation technology by PECVD is described more detailed, and the particularities of the low-frequency direct-plasma PECVD reactor from Centrotherm, mainly used in this work, are outlined.

In **chapter 4**, the results of experiments with a-SiN$_x$:H (SiN$_x$) are presented. SiN$_x$ was solely deposited using the low-frequency Centrotherm system, in contrast to SiC$_x$ (Chapter 5).

Regarding deposition parameters, the influence of the gas flow ratio and wafer position in the horizontal boat position on the passivation and optical properties as deposited and after simulated contact co-firing were investigated.

As an approach to potentially reduce the costs of PECVD deposition by using cheaper precursor gases, the effect of a variation of the purity grade of ammonia used for the SiN$_x$ depositions was investigated on lifetime samples as well as solar cells. Finally, the long-term stability of the encapsulated solar cells was tested by temperature variation cycling as no clear difference between the different purity grades was detectable on the cell level between ammonia purity grades N50 (UHP), N36 and N20 (industrial grade, 99% purity).

Chapter 5 presents the experiments and results of a-SiC$_x$:H (SiC$_x$) depositions from methane and silane by both 13.56 MHz high-frequency and 40 kHz low-frequency PECVD, as well as experiments with a-SiC$_x$N$_y$:H (SiC$_x$N$_y$) deposited by low-frequency PECVD from methane, ammonia and silane.

Resulting films were characterized by QSSPC lifetime measurements and thickness and refractive index evaluated by spectroscopic ellipsometry. The low-frequency deposited samples were later also subject to FTIR measurements to investigate their structure and chemical composition.

The best passivation of τ_{eff}=4 ms at 10^{15} cm^{-3} / S$_{eff}$ <4 cm/s for as-deposited SiC$_x$ on n-type FZ-Si (p-type: τ_{eff}=2.7 ms at 10^{15} cm^{-3} / S$_{eff}$ <5 cm/s) was found for layers fabricated by the high-frequency PECVD with highest Si content (methane/silane gas flow ratio GFR=1), i.e. these layers were stoichiometrically close to amorphous silicon, with the passivation quality continuously decreasing with Si content. This trend changed after firing, with the highest effective lifetime on n-type FZ-Si of 142 μs (S$_{eff}$ <100 cm/s) being reached at a GFR of 6, whereas the passivation of the 4 ms sample almost completely degraded to 4 μs, which is in accordance with a-Si:H results, not able to withstand firing.

SiC$_x$ from low-frequency PECVD gave lower passivation for the as-deposited layers with a maximum of τ_{eff}=400 μs at 10^{15} cm^{-3} / S$_{eff}$ <20 cm/s on p-type Cz-Si, but the best passivation after firing of τ_{eff}=220 μs at 10^{15} cm^{-3} - S$_{eff}$ <30 cm/s is better than for the films deposited by high-frequency PECVD within this work, and equals the best passivation by intrinsic SiC$_x$ reported in [Martin03] and [Ferre08].

Zusammenfassung

Kapitel 1 gibt einen einleitenden Überblick des aktuellen Stands der Photovoltaik, mit Schwerpunkt auf der auf kristallinem Silizium (c-Si) basierenden Technologie. Ein wesentlicher Beitrag zur Reduktion der Stromgestehungskosten ist auf Produktionsebene der Solarmodule vor allem durch reduzierten Siliziumverbrauch durch den Einsatz dünnerer Wafer und/oder den Einsatz billigeren Rohsiliziums ("Solar Grade Silizium") zu erwarten. Die Nutzung des größten Potentials durch Kombination beider Faktoren hängt vor allem anderen von der Verfügbarkeit guter industriell umsetzbarer Oberflächenpassivierungs- methoden ab-dann spielt bei Waferdicken von 30 μm statt der aktuell üblichen 150-200 μm die Materialqualität fast keine Rolle mehr. Dies liefert die Motivation für diese Arbeit.

Kapitel 2 gibt einen Überblick der theoretischen Grundlagen von Oberflächenpassivierung und Antireflexschichten und beschreibt die Methoden und Messgeräte, die zur Charakterisierung der im Rahmen dieser Arbeit hergestellten Schichten verwendet wurden.

Obgleich Oberflächenpassivierung materialunabhängig durch die effektive Oberflächenrekombinationsgeschwindigkeit S_{eff} quantifiziert wird, kann dieser Parameter nicht direkt gemessen werden. Stattdessen liefern die in dieser Arbeit ausgeführten Lebensdauermessungen mittels QSSPC und μPCD die effektive Minoritätsladungsträgerlebensdauer τ_{eff}. Mit geeigneten vereinfachenden Annahmen kann ein oberes Limit für S_{eff} einzig aus τ_{eff} und der Probendicke berechnet werden. Wie im Kapitel 2.2.1 gezeigt, ist der aus dieser Vereinfachung resultierende Fehler für gute Oberflächenpassivierungsschichten nicht vernachlässigbar. Er ist jedoch akzeptabel, da in der Praxis der Schwerpunkt auf dem relativen Vergleich verschiedener Passivierungsschichten liegt.

Während die μPCD zur Bestimmung der räumlichen Verteilung der Passivierungsqualität über die gesamte Probe verwendet wurde, erfolgten anschliessend QSSPC-Messungen um absolute Werte der besten Bereiche zu bestimmen, die mit Literaturwerten verglichen werden können, da die QSSPC zum etablierten Standard in der c-Si Photovoltaik geworden ist.

Der Brechungsindex und die Dicke der untersuchten Schichten wurden mit spektroskopischer Ellipsometrie gemessen, und in einigen Fällen wurde die chemische Zusammensetzung mit Fouriertransformierter Infrarotspektroskopie (FTIR) untersucht, um Zusammenhänge mit der Oberflächenpassivierungsqualität und den optischen Eigenschaften aufzuspüren.

Kapitel 3 erklärt den Unterschied zwischen Aufwachsen und Abscheiden als Methoden, einen Film auf einem Substrat zu erhalten, und gibt Beispiele von Dielektrika und ihren Herstellungsmethoden, die in der aktuellen c-Si Solarzellentechnologie am gebräuchlichsten sind, mit Schwerpunkt auf SiO_x und SiN_x aus thermischer Oxidation beziehungsweise plasmaunterstützter chemischer Abscheidung aus der Gasphase (PECVD, vom englischen "Plasma Enhanced Chemical Vapor Deposition").

Anschließend wird die Herstellung dünner Filme mittels PECVD detaillierter beschrieben, und Besonderheiten des Niederfrequenz-Direktplasma PECVD Reaktors von Centrotherm, der in dieser Arbeit überwiegend benutzt wurde, werden dargelegt.

In **Kapitel 4** werden die Ergebnisse von Experimenten mit a-SiN_x:H (SiN_x) präsentiert. SiN_x wurde ausschließlich mit dem Niederfrequenzsystem von Centrotherm abgeschieden, im Gegensatz zu SiC_x (Kapitel 5).

An Abscheidungsparametern wurden der Einfluss des Gasflussverhältnisses sowie der Waferposition im Prozessboot während der Abscheidung auf die optischen und Passivierungseigenschaften der Schichten untersucht, sowohl nach der Abscheidung als auch nach einem anschließenden simulierten Kontaktfeuerschritt untersucht.

Die mit SiN_x gemessene Oberflächenpassivierungsqualität ist nach Kenntnis des Autors die bisher höchste für ein Niederfrequenz-PECVD-System veröffentlichte, und liegt nur geringfügig unter den besten veröffentlichten Werten für Hochfrequenz-PECVD [Kerr03]. Dies steht scheinbar im Widerspruch zu Veröffentlichungen, die schlechtere Oberflächenpassivierungsqualitäten für Niederfrequenz-PECVD-Systeme aufgrund des durch Ionenbombardement verursachten Oberflächenschadens gefunden haben. Es wird eine wahrscheinliche Erklärung dafür geliefert, warum mit der verwendeten Niederfrequenz-PECVD eine ähnlich gute Passivierungs- qualität erzielt werden kann wie mit einer Hochfrequenz-PECVD.

Als einen Ansatz zur möglichen Kostensenkung der PECVD-Abscheidungen durch die Verwendung kostengünstigerer Prozessgase wurde der Effekt einer Reinheitsgradvariation des zur SiN_x-Abscheidung verwendeten Ammoniak untersucht, sowohl auf Lebensdauerproben als auch in Solarzellen. Abschließend wurde die Langzeitstabilität der in Module einlaminierten Solarzellen durch eine Temperaturwechselprüfung nach Industrienormvorgabe getestet, da auf Zellebene kein merklicher Unterschied zwischen den Reinheitsgraden N50 ("UHP", d.h. hochrein) sowie N36 und N20 (Industriequalität, 99% rein) festzustellen war.

Kapitel 5 präsentiert die Experimente und Ergebnisse von a-SiC_x:H (SiC_x) Abscheidungen mit Methan und Silan, sowohl mittels 13.56 MHz Hochfrequenz- als auch 40 kHz Niederfrequenz-PECVD, sowie Experimente mit a-SiC_xN_y:H (SiC_xN_y), abgeschieden mittels Niederfrequenz-PECVD mit Methan, Ammoniak und Silan.

Die resultierenden Filme wurden durch QSSPC-Lebensdauermessungen charakterisiert, und Schichtdicke und Brechungsindex mit spektroskopischer Ellipsometrie bestimmt. An den mit Niederfrequenz abgeschiedenen Proben wurden später auch FTIR-Messungen durchgeführt, um deren Struktur und chemische

Zusammensetzung zu untersuchen.

Die beste Oberflächenpassivierung von τ_{eff}=4 ms bei 10^{15} cm^{-3} / S_{eff} <4 cm/s für SiC$_x$ wie-abgeschieden auf n-Typ FZ-Si (p-Typ: τ_{eff}=2.7 ms bei 10^{15} cm^{-3} / S_{eff} <5 cm/s) wurde für Schichten gefunden, die mittels Hochfrequenz-PECVD und dem höchsten Si-Gehalt abgeschieden worden waren (Methan/Silan Gasflussverhältnis GFR=1), d.h. diese Schichten waren stöchiometrisch ähnlich zu amorphem Silizium, während die Passivierungsqualität kontinuierlich mit dem Si-gehalt abnahm. Dieser Trend änderte sich nach dem Feuern, mit einer höchsten effektiven Lebensdauer auf n-Typ FZ-Si von 140 μs (S_{eff} <100 cm/s) bei einem Gasflussverhältnis von 6, während die Passivierung der 4 ms Probe fast vollständig auf 4 μs degradierte, was mit Ergebnissen für a-Si:H übereinstimmt, das ebenfalls nicht feuerstabil ist.

SiC$_x$ mittels Niederfrequenz-PECVD ergab geringere Oberflächenpassivierung für die Schichten vor Feuern, mit einem Maximum von τ_{eff}=400 μs bei 10^{15} cm^{-3} / S_{eff} <20 cm/s auf p-Typ Cz-Si, während die beste Passivierung nach Feuern von τ_{eff}=220 μs bei 10^{15} cm^{-3} - S_{eff} <30 cm/s besser ist als die der mittels Hochfrequenz-PECVD im Rahmen dieser Arbeit abgeschiedenen Filme, und auf gleichem Niveau liegt wie die beste von [Martin03] und [Ferre08] mit intrinsischem SiC$_x$ erreichte Passiverung.

Publications

J. Libal, R. Petres, T. Buck, R.Kopecek, G. Hahn, R. Ferré, M. Vetter, I. Martín, K. Wambach, I. Röver, P.Fath, *N-Type Multicrystalline Silicon Solar Cells - BBr$_3$ Diffusion and Passivation of p$^+$ diffused Silicon Surfaces*, Proc. 20th EU PVSEC, Barcelona, 793-796, 2005

R. Petres, J. Libal, R.Kopecek, M. Vetter, R. Ferré, I. Martín, D. Borchert, I. Röver, K. Wambach, P.Fath, *Passivation of p$^+$ Surfaces by PECVD Silicon Carbide Films - A Promising Method for Industrial Silicon Solar Cell Applications*, Proc. 15th PVSEC, Shanghai, 129-130, 2005

R. Kopecek, T. Buck, J. Libal, R. Petres, I. Röver, K. Wambach, R. Kinderman, L. J. Geerligs, P. Fath, *Large Area N-Type Multicrystalline Silicon Solar Cells with B-Emitter: Efficiencies Exceeding 14%*, Proc. 15th PVSEC, Shanghai, 883-884, 2005

J. Libal, R. Petres, R.Kopecek, G. Hahn, M. Vetter, I. Röver, K. Wambach, P.Fath, *N-Type Multicrystalline Silicon Solar Cells: PERC Design for High Efficiency*, Proc. 15th PVSEC, Shanghai, 969-970, 2005

R. Petres, J. Libal, T. Buck, R.Kopecek, M. Vetter, R. Ferré, I. Martín, D. Borchert, P.Fath, *Improvements in the Passivation of p$^+$-Si Surfaces by PECVD Silicon Carbide Films*, Proc. 4th WCPEC, Waikoloa, 1012-1015, 2006

T. Buck, R.Kopecek, J. Libal, R. Petres, K. Peter, I. Röver, K. Wambach, L. J. Geerligs, E. Wefringhaus, P. Fath, *Large Area Screen Printed N-Type mc-Si Solar Cells with B-Emitter:Efficiencies close to 15% and Innovative Module Interconnection*, Proc. 4th WCPEC, Waikoloa, 1060-1063, 2006

M. Vetter, R. Ferré, I. Martín, P. Ortega, R. Alcubilla, R. Petres, J. Libal, R.Kopecek,*Investigation of the Surface Passivation of p$^+$-Type Si Emitters by PECVD Silicon Carbide Films*, Proc. 4th WCPEC, Waikoloa, 1271-1274, 2006

L. J. Geerligs, C.J.J. Tool, R. Kinderman, I. Röver, K. Wambach, R. Kopecek, T. Buck, J. Libal, R. Petres, P. Fath, P. Sánchez-Friera, J. Alonso, M. Acciarri, S. Binetti, S. Pizzini, *N-Type Solar Grade Silicon for Efficient P$^+$N Solar Cells: Overview and Main Results of the EC NESSI Project*, Proc. 21st EU PVSEC, Dresden, 566-569, 2006

E. Wefringhaus, R. Petres, A. Helfricht, A. Remgérd, K. Lundahl, *Electroless Silver Plating of Screen Printed Front Grid Fingers as a Tool for Enhancement of Solar Cell Efficiency*, Proc. 22nd EU PVSEC, Milano, 1050-1052, 2007

J. Libal, M. Acciarri, S. Binetti, R. Kopecek, R. Petres, C. Knopf, K. Wambach, *Effect of Extended Defects on the Electrical Properties of Compensated Solar Grade Multicrystalline Silicon*, Proc. 22nd EU PVSEC, Milano, 1124-1129, 2007

K. Peter, R. Kopecek, R. Petres, P. Fath, E. Wefringhaus, *Long Term Perspective for Photovoltaic R&D Activities in Konstanz*, Proc. 22nd EU PVSEC, Milano, 3519-3520, 2007

S. Riegel, B. Raabe, R. Petres, S. Dixit, L. Zhou, G. Hahn, *Towards Higher Efficiencies for Crystalline Silicon Solar Cells Using SiC Layers*, Proc. 23rd EU PVSEC, Valencia, 1604-1607, 2008

J. Libal, S. Novaglia, M. Acciarri, S. Binetti, R. Petres, J. Arumughan, R. Kopecek, A. Prokopenko, *Effect of compensation and of metallic impurities on the electrical properties of Cz-grown solar grade silicon*, J. Appl. Phys. **104**, 104507 (2008)

R. Petres, V. D. Mihailetchi, J. Maier, S. Keller, T. Pernau, P. Fath, *Silicon Surface Passivation by Industrial Low Frequency PECVD Films - Properties and Performance of SiC_x and SiO_xN_y*, Proc. 24th EU PVSEC, Hamburg, 1613-1616, 2009

A. Madec, H. Chevrel, R. Petres, S. Eisert, K. Peter, *Impact of NH_3 Grade Used for PECVD of a-SiN_x:H on Silicon Solar Cell Performance*, Proc. 24th EU PVSEC, Hamburg, 1315-1317, 2009

A. Madec, H. Chevrel, R. Petres, S. Eisert, K. Peter, *ARC Deposition with Various NH_3 Grades: Impact on c-Si Solar Cell Performance*, Proc. 5th WCPEC, Valencia, 2010 (in print)

Acknowledgements

Many people and institutions have played a role in the making of this thesis. Following up, I will try to mention as many of them as possible, while their order shall not correspond to an accurate "importance ranking", which seems as difficult to come up with as a complete list of all names.

I wish to thank Prof. Dr. Ernst Bucher for accepting me as a Ph.D. student even after becoming emeritus and for providing numerous literature sources. I wish to be similarly active and passionate about my work years after the official retirement.

Prof. Dr. Johannes Boneberg for accepting the co-assessment of this thesis.

Dr. Sara Olibet for proof-reading and correcting this thesis, providing a lot of helpful input at any time of the day and week.

Dr. Valentin Mihailetchi for insightful scientific discussions as well as help with simulations and DoE evaluation software.

Dr. Jayaprasad Arumughan and Amir Dastgheib-Shirazi for help with measurements and processing at literally any time of the day and week, as well as their long-lasting friendship.

Dr. Kristian Peter, Dr. Radovan Kopecek, Dr. Eckard Wefringhaus and Rudolf Harney for their scientific and moral support and countless discussions, and the nice working atmosphere in the common office.

Stefan Schmitt for contributing the interesting results of his Bachelor thesis, and his diligence and reliability.

All other people at ISC Konstanz and the PV research group at the University of Konstanz for support with numerous measurements, processing and many scientific and non-scientific discussions as well as the nice working atmosphere.

Dr. Michael Vetter, Dr. Isidro Martin and Dr. Rafel Ferre, then at the Universitat Politecnica de Catalunya, for the warm welcome in Barcelona, help with processing and characterization of silicon carbide samples, and the nice working atmosphere in their research group.

Dr. Dietmar Borchert and Dr. Markus Rinio at the Fraunhofer ISE Gelsenkirchen for their cooperation and support with silicon carbide depositions.

Die VDM Verlagsservicegesellschaft sucht für wissenschaftliche Verlage abgeschlossene und herausragende

Dissertationen, Habilitationen, Diplomarbeiten, Master Theses, Magisterarbeiten usw.

für die kostenlose Publikation als Fachbuch.

Sie verfügen über eine Arbeit, die hohen inhaltlichen und formalen Ansprüchen genügt, und haben Interesse an einer honorarvergüteten Publikation?

Dann senden Sie bitte erste Informationen über sich und Ihre Arbeit per Email an *info@vdm-vsg.de*.

Sie erhalten kurzfristig unser Feedback!

VDM Verlagsservicegesellschaft mbH
Dudweiler Landstr. 99   Telefon +49 681 3720 174
D - 66123 Saarbrücken   Fax   +49 681 3720 1749

www.vdm-vsg.de

Die VDM Verlagsservicegesellschaft mbH vertritt

Printed by Books on Demand GmbH, Norderstedt / Germany